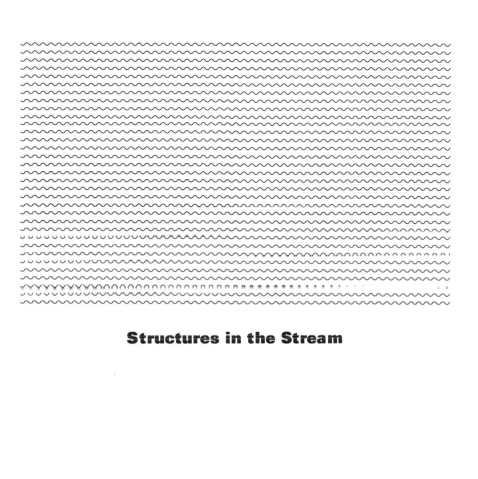

Structures in the Stream

American Studies Series

WILLIAM H. GOETZMANN, EDITOR

STRUCTURES IN THE STREAM

Water, Science, and the Rise of the U.S. Army Corps of Engineers

by Todd Shallat

UNIVERSITY OF TEXAS PRESS  AUSTIN

Requests for permission to reproduce material
from this work should be sent to:
Permissions, University of Texas Press,
Box 7819   Austin, TX 78713-7819

♾ The paper used in this publication meets
the minimum requirements of American
National Standard for Information Sciences—
Permanence of Paper for Printed Library Materials,
ANSI Z39.48-1984.

LIBRARY OF CONGRESS
CATALOGING-IN-PUBLICATION DATA

Shallat, Todd A.
Structures in the stream : water, science, and the rise of the U.S.
Army Corps of Engineers / Todd Shallat. — 1st ed.
p.   cm. — (American Studies series)
Includes bibliographical references and index.
ISBN 0-292-77679-9
1. Water resources development—Social
aspects—United States—History.   2. Water resources develop-
ment—Political aspects—United States—History.   3. United States.
Army. Corps of Engineers—History.   4. Hydraulic engineering—
United States—History.   5. United States—Civilization—French
influences.   I. Title.   II. Series.
TC423.S49   1994
333.91′15′0973—DC20   93-42972

West Point engineers have not their superiors anywhere;
they know all that can be known of their science;
and so, since they conceived that they can fetter and handcuff
that [Mississippi] river and boss him, it is but wisdom
for the unscientific man to keep still, lie low, and wait till they do it.

—Mark Twain,
Life on the Mississippi
1883

The total domination of nature
inevitably entails a domination of people
by the techniques of domination.

—André Gorz,
Ecology in Politics
1980

Contents

Preface

I live in Idaho on a tributary of the Snake where the U.S. Army Corps of Engineers is widely misunderstood and despised. No matter that a Corps dam just up the river protects the valley from flooding. No matter that we southern Idahoans tap federal projects for more gallons of water per capita than Americans in any other part of the nation, or that, thanks in part to the Corps, our region enjoys some of America's cheapest hydropower. The Corps remains, for many, the epitome of an entrenched bureaucracy, a bulldozer out of control.

Attacked by salmon advocates, kayakers, irrigators, the port of Lewiston, and others, the Corps keeps a low profile in Idaho, downplaying its critical role as the architect of a Snake-Columbia River dam and canal network that, just a generation ago, was praised as one of the great engineering feats of our times. Recently I supervised a contract history of a flood project in that remarkable Snake-Columbia system. The historian, my graduate student, explained that the Corps had "helped" developers alter the flood plain to accommodate a housing project. "We don't 'help,'" a Corps engineer wrote in the margin. "We provide 'technical assistance,'" That tiny distinction spoke volumes about the reviewer's sense of himself as an impartial technician. His point was that the Corps does not take it upon itself to promote projects, discourage projects, or ally itself with a faction in a local dispute. Executing the mandates of Congress, the Corps merely responds.

Structures in the Stream is my reply to that agency spokesperson and also my way of thinking through the connection between science, technology, and political power. A probe into the origins of federal engineering and science, it follows the rise of army construction from Jeffersonian times through the boom era of waterway projects after the Civil War. The book is not meant to be an environmental history, although ideas about the human domination of nature intersect with the story line. Instead this book is about a European building tradition that matured in the early Republic and survives into our times. A study of engineering ideology, the book is also about the origins of the presumption that a government armed with science could solve great mysteries of nature and reorder American life.

Because some of the language used in the book has fallen out of favor, a word of explanation is due. I use "America" and "Americans" in the same prideful and provincial way the first citizens of the United States of America once used those words—to refer exclusively to themselves, as if people south and north of the border were invisible or inconsequential. My rationale is that the culture in question deserves to be understood in its own terms, especially since the parochial "American-ness" of the early Republic is an important theme of the book. Likewise, I use masculine pronouns and suffixes if the people in question were, like army builders in the early Republic, an exclusively male group. Effects of that maleness on the engineer psyche are considered in Chapter 5.

My interest in these subjects goes back to the Carter and Reagan years when I wrote water history as a contractor with the Corps' Historical Division in Washington, D.C. (now the Office of History at Fort Belvoir, Virginia). Flattering or not, my work was always encouraged. I owe all the Corps historians a towering debt—especially Bill Baldwin, Dale Floyd, John Greenwood, Martin Gordon, Charles Hendricks, Leland Johnson, Martin Reuss, Frank Schubert, Barry Sude, and Paul Walker. Dr. Reuss, my secret weapon, sparkled with so much vital information that I can no longer separate our ideas.

Friends and mentors who helped me grow as a scholar are part of all that I write. I profited immeasurably from time spent with Carl Harris and Robert Kelley, patient men who introduced me to historical writing. Over the years I have also benefited from lively conversations with John Beeman, Brooke Hindle, Suellen Hoy, James Martin, Darwin Stapleton, Jeffrey Stine, Michael Robinson, Howard Rosen, and Philip Shiman. At Carnegie-Mellon University the amazing Joel Tarr, a probing mind, moved me with a passion for applied social science and directed the dissertation that was an early draft of this book.

At Boise State University the people who endure me daily have been there when I needed them most. Milton Small read everything, offering comment. Errol Jones and Robert Sims sustained me through hard times and endorsed my applications for grants. Special thanks also to Michael Zirinsky, Amy

Stahl, Nicholas Casner, Marilyn Paterson, Karen Kelsch, Denise Fitelson-Nelis, Madeline Buckendorf, the History Department, my questioning graduate students, and the university librarians. Generous support from the College of Social Sciences and Public Affairs helped pay for maps and illustrations. The Corps' Office of History also helped purchase illustrations.

Portions of the study have been adapted for journals. I especially enjoyed my work with *Technology and Culture*, *Natural Resources Journal*, and *The Public Historian*.

Finally, no one has been closer to this project than my wife, Nancy Shallat. Only Nancy really knows the look on my face when friends ask me what I did on my summer vacation. To her, and our son Tony, I dedicate this book.

A Nation Builder

The American faith in progress through science and engineering has often rested on the idea that government, as it strived for a more perfect union, would be able to perfect nature as well. "What more noble or holy duty remains?" asked the Whig politician John C. Spencer, imploring Congress to complete what God himself had begun. Spencer, writing in 1847, called the United States a boundless hydrographical system. Apart from a few sand bars and shoals, the rivers and lakes were "great conduits" of navigation, a five-thousand-mile circuit from New England to the West and the Gulf. Water bound the republic, fusing the states, making America "one government, . . . a union." [1] To preserve that union, however, Spencer advocated a large and steady investment in river and harbor construction. Americans, he insisted, would also have to invest in planners and waterway experts. Impartial and scientific, they would elevate engineering, ensuring that the nation spent money on just and reasonable plans.

This book examines the rise and spreading influence of those scientific experts, their early technologies, their political agenda, and the mark they left on the union through their embattled public works organization, the United States Army Corps of Engineers.

The Corps has been called America's preeminent engineering organization. A nation builder. A bureaucratic superstar. Also a public enemy, a diligent destroyer of wetlands, a military aristocracy, a lobby that can't be licked. [2] Created in 1802, the organization began as a war academy and fort-building agency, but the idea of an engineer corps stretched more than a century back to the time of Louis XIV when an elite and scientific force of government planners modernized the kingdom through highways, waterways, aqueducts, and other spectacular projects. Americans have long since admired that French tradition of public works—and despised it as well. To the extent that our constitutional democracy remains a product of Great Britain—egalitarian, capitalistic, suspicious of government experts and peacetime armies—the Corps of Engineers has often worked outside the mainstream of American culture. Thus, the two hundred years of debate over how the nation should develop its water resources has often been a dispute over the alien and elitist traditions of the nation builders themselves.

Why return to this old debate? Because bureaucracy, a Freudian child, matures through stages of conflict, its character marked in formative years. During the turbulent first century of federal aid to engineering projects, about 1789 to 1880, the Corps of Engineers emerged as the nation's largest and most powerful water development agency, a military-led organization that survived attacks on the army by managing three kinds of disputes. First was a cultural and professional contest that distanced the French engineer-scholar from British and American builders who learned their craft in the field. Second was a technological conflict over the practical value of science. Third was a political struggle over bureaucratic power. Resisting the British example of the self-made builder-mechanic in a freewheeling capitalist system, the Corps leadership, borrowing chiefly from the French model, favored a planned economy where the army guided construction and science was the methodical tool of a rational, centralized state. That vision of a planned republic has always met stiff opposition. In a nation devoted to private enterprise and community rights, the three angry lines of anti-Corps, antiscience, antibureaucratic resistance have fractured the building professions and frustrated the army campaign to federalize public works.

The Corps' problems began with resistance to science and militarism, a cultural dispute. Fueled by warfare between two great powers in Europe, the conflict fed on the friction between French-style government planning and a British preference for free competition among bold entrepreneurs. Engineers stood at a crossroads between two European ideals. "I do not know how it has happened," wrote Chief Engineer Jonathan Williams, reporting from Corps headquarters at West Point in 1802; "I cannot find any full English Idea [for] what the French give to the profession." The French builders were "men of science." Schooled for government service, they rejected the British example of the craftsman or lowborn mechanic. "To be merely an engineer, an inventor, a maker or Director of Engineers is one thing," Williams insisted; "to be an officer *du génie* is another." [3]

The ideal of the engineer-scholar left the new Corps of Engineers with French inclinations and values: a talent for planning and applied mathematics, a flair for monumental construction that empowered the nation-state. At the U.S. Military Academy at Westpoint, the Corps maintained a library of French publications. Professors became translators. In 1817 the Frenchman Claudius Crozet applied geometry to canal design in America's first scientific course on construction. French architecture dominated Corps fortifications. French jetties and breakwaters became models for American projects. French standardization drove the army campaign for musket and rifle production through interchangeable parts.

France also sold the American army on the promise of national planning. After 1824, when Congress started to finance water construction, an army board of engineer-planners, headed by one of Napoléon's best engineers, looked abroad for the administrative reforms, river technologies, and scientific innovations that set a high standard for federal public works.

So pervasive was the Gallic tradition that the army ordinance board rejected British equipment for inferior French designs. Yet a small group of army technicians, the surveyors, applauded British ideas. Since colonial times the British army had used its surveyors and scouts to investigate navigation.

The United States Military Academy at West Point, New York, around 1854. Established in 1802, the military academy was Corps headquarters until the chief engineer moved to Washington after the War of 1812. Courtesy of the Library of Congress.

Geographer Thomas Hutchins, the American-born British officer who later founded the 1785 federal land survey, studied West Florida and Louisiana and profited from the demand for scientific information about the Ohio-Mississippi system. Soon the frontier army was pushing for river improvements and far-reaching roads and canals. In reports by Lewis and Clark, Zebulon Pike, Andrew Ellicott, and others, surveying and science converged in a literature on the strategic importance of water communication. Army maps and reports became aids to commerce that marked the defensible limits of territorial expansion. For Maj. Stephen H. Long, a nineteenth-century topographical engineer, access to navigable rivers was a military necessity that determined the worth of federal lands. Defining the strategic perimeter of American civilization, Long found desert in the West and wasteland in the North that were barriers to steamboat navigation. Long lent a military, sometimes circular logic to federal water projects: if a waterway was navigable it was important and worth defending; once fortified, the channel was worth improving to facilitate defense.

The growing importance of waterway surveying led to the second conflict—a technological dispute. It began with some confusing semantics. What the military called "science" was an organizational philosophy with political implications that were always controversial. Although a literature on engineering now insists that the goal of science ("understanding nature," said historian Melvin Kranzberg) is quite different from the goal of technology ("making useful things"), the clarity of that distinction was lost in the West Pointer's world.[4] Army waterway science was order and classification. Rational and precise, it spelled out the theory or natural laws that reduced river construction to a regimen of standardized steps. Science was also politics in the 19th century army. As the study of rivers and harbors moved from classrooms to the field, the Corps became a spokesman for massive and complex "scientific" projects: canals, dams, aqueducts, lighthouses, levees, breakwaters, and ports—improvements that seemed extravagant to many American builders. Corps science set a grandiose standard for public construction. It justified the cost and complexity of river improvement programs that were targets of the rising resistance to federal public works.

Disputes over the meaning and value of science stirred a deeply ingrained opposition to military bureaucracy—a third conflict. In Jacksonian times the attack on the Corps drew strength from a bitter campaign to abolish the school at West Point. Critics denounced West Point and the Corps as tools of centralization. Militiamen said West Pointers monopolized high rank in the army. Others said Corps engineering violated the rights of the states. Scandals and bad engineering fueled the angry resistance, and in 1836, after a series of construction fiascos, the Corps was publicly censured. Two years later the chief engineer was under investigation for fraud. Waterway programs survived, but Congress restricted the Corps and suspended the Whig campaign for a unified public "system" of army-built water projects. Even after the Civil War, when farmers looked to the rivers to break the grip of the railroads and every town with a wharf was demanding federal aid, a populist attack on the Corps split the engineering profession. Condemning the engineer "despot" and vilifying his science, the protest, peaking in the 1880s, undercut army attempts to consolidate federalism by extending the five-thousand mile circuit of nature's near perfect plan.

That the waterway program continued to grow despite the fear of militarism said more about logrolling in Congress than the nation's goodwill toward the Corps. As Louisiana led the campaign for federal flood protection and the upper Mississippi Valley pressed for deep-water shipping channels, Congress, uniting South and Midwest, passed veto-proof legislation with something for every region. "No outsider pleads for the public interest," said Professor Albert B. Hart of Harvard, reporting in 1888 on the "extravagance" of the pork-barrel construction.[5]

Hart's call for reform focused on the fragmentation of the waterway planning process. Working through harbor boards, canal boards, levee commissions, debris commissions, and bridge-inspection teams, the Corps responded to local pressure by promising grand construction for a fraction of the probable cost. Perhaps the most outrageous example was the plan the Corps sold to Congress in 1887. Blocked by President Cleveland, the first successful veto of a waterway bill since Buchanan's time, it pledged $10 million to 439 projects. Only 8 of the projects were fully funded. The real cost, according to Hart, was at least $200 million. When critics objected to the outright deception, Congress—driven by Theodore E. Burton, the powerful chair of the House Rivers and Harbors Committee—modified waterway spending by adding "authorizations" that, in effect, forced the government to make regular payments on unfinished jobs. Thus in 1899 a $15.8 million bill promised another $21.5 million in authorizations. The bill passed the House by a landslide vote of 160 to 7. Yet an editorial in *The Nation* denounced the proposed legislation, calling it "the most extravagant rivers and harbors bill ever known."[6]

As critics linked the Corps to the waste and corruption in Congress, a new school of efficiency experts switched places with army builders in the scientific planning crusade. Once it had been the soldiers who preached the loudest for order through centralization. By the 1880s, however, the campaign for top-down waterway planning was often an attack on the Corps. "The

Critics have long denounced the "pork-barrel" ties that bind the Corps to the Senate and House. Courtesy of the *Arkansas Democratic Gazette* and George Fisher.

practice and method of the Engineer Corps are a disgrace to the nation," said George Y. Wisner, a contributor to *Engineering Magazine* in 1892. "Their inefficiency is beyond question," yet, Wisner explained, the "members of Congress . . . are averse to taking any official action to place public improvements on a better business basis, for the reason that their constituents expect them to secure the usual appropriation."[7] Again in 1907 a reformer maintained that the Corps, unlike the new U.S. Bureau of Reclamation, failed to take a "large outlook" on waterway development: "there is no connection between any two projects and therefore no general benefit to the nation."[8] For the early conservation movement, and for conservationists even today, the critique of the Corps helps explain why the United States remains one of the very few industrialized countries without a central planning authority or an intelligible national plan.[9]

Was the Corps, then, partly to blame for the failure of waterway planning? Samuel P. Hayes thought so. In *Conservation and the Gospel of Efficiency* (1959), an important book on water and land politics during the Progressive Era, Hayes showed how a bullheaded, back-stabbing Corps frustrated frequent attempts to change the way Congress did business. Few students of waterway planning have strayed far from Hayes's critique. Historian Otis L. Graham said the Corps was exacting revenge on the water reformers when it conspired to "murder" a brave experiment in New Deal planning, the National Resources Planning Board.[10] For Elmo Richardson, a student of dams and national parks during the Truman-Eisenhower era, the Corps, a myopic giant, was chained to the power brokers and land corporations who blocked conservation reform in the Pacific Northwest.[11]

Although the builders have been quick to dismiss these kinds of historical studies as cynicism from academe, the modern scholarship builds on a Jeffersonian dread of bureaucratic corruption, a long-standing grassroots critique. Rooted in an ancient terror of the peacetime professional soldier, aggravated by the military-civilian rift through the engineering profession and a history of intergovernmental feuds, the critique, historically, has fixed on the "power elite" or "power fraternity" that one critic has called "the most powerful and most pervasive lobby in Washington"—the Corps leadership trained at West Point.[12] That gray line of army tradition has been "rigid" and "highly resistant to change," said Arthur E. Morgan, a former chairman of the Tennessee Valley Authority (TVA).[13] But elsewhere the dam building West Pointers seemed all too eager for new kinds of assignments. "The Corps," wrote Fred Powledge in *Water* (1982), "is accomplished at what is known as the 'camel's nose' approach, which refers to the notion that if a camel is allowed to poke its nose into a tent, it will not be long before the entire beast is inside."[14] Fiercely competitive, the Corps, author Marc Reisner believed, is "as opportunistic and ruthless an agency as American government has ever seen."[15]

Implicit in the critique is the belief that the Corps is a hidden government that skews engineering and science to serve its fascination for power. "Science," according to a recent book of essays on military research, "has become institutionalized on the basis of the binary schematism of 'true-untrue'— that is, the socially established criterion of truth versus non-truth." Yet army engineering remains a "power-nonpower" world where government presses its own construction agenda, where the demand for secrecy cuts into the flow of scientific information, and where military men defy the first principle of capitalism—the driving preoccupation with cost.[16]

Naturally, the Corps disagrees. Few engineers see a conflict between the values of truth-seeking science and respect for government's power. Said Chief Engineer Frederick J. Clarke, recalling, in 1979, his agency's quick response to the environmental movement: "I think by and large, the people in the Corps realized that our job was to do what we always had done: do what the people of the country wanted."[17]

Clarke was right, of course, and wrong. Wooden obedience to the nation is basic to the West Point concept of military professionalism. Yet no agency is a democracy—certainly not a quasi-military organization that still prides itself as a battle-ready elite. As Francis E. Rourke has explained in his study of program implementation, bureaucrats, especially those who control information, shape the policy process in three possible ways: first, they advise politicians; second, they decide how the programs created by Congress are administered in the field; and third, as technical experts, they make decisions about contracts, worker safety, hirings, firings, and many other vital details. Politicians may have lofty ambitions, but agencies encourage change or doom the government to a policy of inaction. "Administration can make or break a policy," Karl W. Deutsch maintained. "If a policy is to work, or if a leader or party is to have real power, it must have the support of a body of administrative personnel that is loyal and competent enough to give effect to its orders." [18]

In the face of this social science, the Corps has always maintained that good engineering is neutral. One future chief engineer, writing in 1885, said the Corps was "restricted to what has been ordered by the people through its authorized mouthpiece, Congress." [19] A quarter century later the secretary of war agreed that the agency "only executes the will of the people when expressed in legislative form." [20] Likewise in 1977, when the engineer-philosopher Samuel Florman defended the Corps in *Harper's,* he explained that the agency's civil-works directorate, a civilian operation, was "an instrument exquisitely tuned to work the will of the people." [21] Why, then, was the Corps a national scapegoat? Because, said Florman, the agency had been smeared by a regressive environmental movement. Congress tends to agree. "Don't blame the Corps for anything they do," said a member of the Senate Public Works Committee; "blame Congress, because we control the Corps." [22]

Given our national suspicion of Congress, the pride we take in our engineering and science, our love/hate for the West Point tradition, and the stark debate over the power that bureaucracy seems to have on our lives, it is surprising that the written history of the Corps has remained, as historian Alex Roland put it, "sadly lacking." [23] Surveying the neglected field in 1985, Roland found only three notable books. Two were almost thirty years old, and the third, a biographical dictionary, was never published. The standard reference remains Forest Hill's *Roads, Rails and Waterways: The Army Engineers and Early Transportation* (1957). A tribute to the West Point tradition with forty-five pages on navigational works, the book showed how the army elevated construction through dispassionate professionalism, a cold respect for the facts. Thus Hill said nothing about scandals, fiascos, bad engineering, professional dispute, or covert powers of program implementation—subjects also neglected by the many Corps-sponsored division and district histories with patriotic titles like *To Great and Useful Purpose* and *Engineers for the Public Good.* Even Raymond H. Merritt's useful study of the St. Paul District, a book with the words "conflict" and "controversy" in its title, failed to question the agency line about the wall of separation between engineering and policy-making. "Congress," Merritt explained, missing the point about the political weight of Corps expertise, "is the only agency with the right to formulate a national or even regional policy of water resources development."[24]

The paucity of research leaves scholars with shaky assumptions about the politics of engineering and the way our government works. One argument made by defenders and detractors alike is that the Corps was hostile to the conservation ideal of efficiency through scientific management and basin-wide river planning. Actually it was the Corps that pioneered the efficiency-through-science concept. As America's original and most influential planning organization, it bristled at civilian attempts to co-opt army ideas.

Another assumption open to question is that the Corps developed many of the administrative innovations (standardized forms, written guidelines, precise accounting, field offices, clear lines of command) that became a "management model" for the nation's first large corporations.[25] In Jacksonian times, it is true, many West Pointers supervised railroads, but the study of waterways shows that science and regimentation often grated against capitalism. Nor is it likely that the Corps was a significant model for the civil engineering profession. Civilian builders were mostly suspicious of the army's academic training. After 1865, when that kind of training was no longer considered eccentric, the Corps was roundly denounced as a backward organization, and few builders outside the army regarded West Point as a top engineering school.[26]

It was in government, not the private sector, that Corps ideology flourished. There at the elbow of Congress the builders read the mood of the nation: their methods could be conservative, aggressive, flexible, or resolute depending on the personalities in power. As sociologist William F. Ogburn explained in 1933, technologists can only succeed if their society is prepared to deal with the change. When engineers marched ahead of the nation—when the Corps, for example, advocated truss bridges or stone-lined canals—there was often a lapse of time before the public endorsed the invention, and pressed Congress to act. Elsewhere, however, it was Congress that prodded a slow-moving army, spurring innovation. The history of rivers and harbors is rich with examples of both.[27]

In the final analysis, then, the rivers that bound the republic were lines of division as well. Ribbons of commerce, thoroughfares of empire that transmitted federal power, the rivers moved Congress toward planning and bonded engineers to the state. Planning pushed centralization. Fear of the government planner shielded community rights. Since 1802 the weight of our federal system has shifted in both directions—from nationalism to localism, from caution to change.

1

European Antecedents

France civilized the engineer, Britain made him a gifted mechanic, and the new United States profited from both traditions as Americans searched for solutions to hazardous navigation. By the time of the American Revolution, the French engineer had evolved from a soldier to a scholar, from a fort builder to a civil administrator and director of public projects. Meanwhile the British were turning to energetic canal builders who labored for private enterprise and cut through abstractions of science. The Dutch built locks and levees. Italians experimented with flood control. Spaniards pioneered ditch irrigation. Each excelled in an aspect of hydraulic construction, but Americans focused on Britain and France. In the Jeffersonian republic, where the English seemed blunt and pragmatic next to the scholarly French, the distance from Paris to London was the technological span of civil engineering from natural philosophy to practical mechanics, from applied science to art.

One American who came to appreciate the international alternatives was a product of many cultures, a British-born, German-educated descendant of French Huguenots, Benjamin Henry Latrobe. Latrobe is probably best remembered as the architect who rebuilt the U.S. Capitol after the British torched the building in 1814. Engineers know him for the Philadelphia Waterworks (1799–1801), the Susquehanna River improvements (1801–1802), and his plan for the initial phase of the Chesapeake and Delaware Canal (1803–1806). He was also the exception that proved the rule about "two sorts" of American builders. "The first," said Latrobe in a letter to a pupil, were self-proclaimed "architects" who "from travelling or from books have acquired some knowledge of the theory of the art." The second group were

Benjamin Henry Latrobe, architect and engineer, an early advocate of scientific professionalization. Courtesy of the National Portrait Gallery.

"those who know nothing but the practice."[1] Latrobe's hope for America was to see these factions converge.

When the two sorts of builders faced the task of construction in water—aqueducts, dams, canals, harbors, and other hydraulic projects—they drew inspiration from Europe, usually Britain or France. Latrobe was the rare American with experience in both technological worlds. A true cosmopolitan of the scientific Enlightenment, his training began in England in a Moravian school. From age twelve to twenty, 1776 to 1783, he studied on the Continent, principally in German Silesia. As a tourist in Italy and France he sketched public buildings. We know he painted in Paris. Latrobe's study of a Paris hospital resembled his later design for the University of Pennsylvania medical school. Family legend says he attended classes at the University of Leipzig, although there is no surviving record. He may have learned to diagram fortifications in the army of Frederick the Great before returning to family in London in 1783. There he worked about a year in the office of a famous builder, John Smeaton. By 1789 he was a resident engineer under a Smeaton protégé, William Jessop. A lifelong student of hydraulics, Latrobe experimented with pumps and water-powered machinery. He planned freight canals. He mastered the British method of straightening navigation by cutting through bends in rivers. He also acquired a British appreciation for industrious "building artisans," but Latrobe was French in his admiration for builders of breeding and culture, schooled engineers like himself. "The French engineers," he once told President Madison, "have generally, I may say *always*, a better education in the *Science* of their profession."[2]

By the time Latrobe migrated to Virginia in 1795, he was already a fourth-generation American through the Pennsylvania German of his mother's line. Still, critics called him rootless, a "travelling engineer."[3] One day, while dining at Eagle tavern in Richmond, Latrobe heard a revealing story about a French architect with "the most preposterous design" for a house.[4] During the course of conversation Latrobe discovered that he was the Frenchman and the bizarre design was his own. Marked by his accent, distrusted by Federalists but embraced by ardent Francophiles in Thomas Jefferson's circle, Latrobe joined a displaced society of cultured émigrés. His contacts were soldiers, artists, diplomats, and exiled financiers—men like West Point commander Anne-Louis de Tousard, drawing instructor Maximilian Godefroy, rival architects Joseph Mangin and Pierre Bauduy, topographer Guillaume Tell Poussin, and the mysterious engineer who planned the defenses for the Battle of New Orleans, Bartholomy Lafon.

What these Frenchmen had in common, besides America and Latrobe, was a scientific background grounded in formal education. Most had served in the French army. All were fortification experts who found occasional employment as surveyors of public works. Although Latrobe conceded that Frenchmen in uniform were sometimes "pretenders," the properly schooled engineers, graduates of academies like the Ecole polytechnique, were the best-trained builders in Europe.[5] Latrobe developed this theme in an open letter to Madison and the state of Virginia on the planning of roads and canals. "In the question as to employment of an English or French Engineer," Latrobe wrote from Washington, D.C., in 1816, "if the qualifications were equal, the advantage of language might probably decide in favor of the former." An English surveyor might project a level turnpike. A mason who could build an arch might bridge a stream. But structures *in* the stream were more problematic. In Virginia, where rough terrain complicated navigation projects, a builder needed "industry," "experience," and "scientific acquaintance with that branch of civil engineering that relates to Canals." The man must also have "a vigorous constitution capable of undergoing great fatigue." If there existed in all of America a truly qualified engineer, and Latrobe was unsure, he was probably a European officer familiar with hydraulics. "In this climate," Latrobe insisted, "a Frenchman has the advantage."[6]

Virginia, for the time being, rejected Latrobe's advice, although the state did hire a French engineer in 1823. Meanwhile the British still dominated the transatlantic transfer of construction technology. French officers remained aliens outside that tradition. America's first canal and turnpike companies preferred to hire from London. Britain also provided the laborers, machines, tools, and flight capital that fueled these corporations. And because both cultures admired hands-on men of action, the Anglo-American builders had little patience for science. In 1808 a Boston journal for inventors called America "repugnant to theory." [7]

Nearly two centuries later the bulk of historical writing still centers on craft tradition. Stressing British mechanics, the literature on Yankee construction slights the French and their science. Melvin Kranzberg said prerevolutionary French technologies were "aristocratic" and therefore "ill-suited" to American culture.[8] Elting Morison saw an American technology emerging from the fingertip knowledge of the antebellum craftsman. "There were not available at that time many general ideas or theoretical considerations that would help in building or making new things," Morison explained. "Men derived useful procedures—rules of thumb—from practical experience." [9] Thus, according to Morison, Kranzberg, Eugene Ferguson, and others, the greatest builders of the age began with trial and error. The model engineer is a civilian composite—not a Latrobe but a John B. Jervis, a Benjamin Wright, an untutored builder with expedient methods, an inventive mechanic.[10]

But what of hydraulic construction, the product of both craftsmanship and science, a popular subject of scholarly investigation in continental Europe? Here there remained a division within civil engineering. During the early years of navigation improvement, as the great powers vied for influence in rising America, an engineer's approach to hydraulics could vary according to many factors—schooling, place of employment, exposure to science, and cultural orientation. As Latrobe had suggested, two sorts of builders emerged. One, largely British, met a civilian demand for skilled mechanics. The other was military, scientific, and true to the French idea that a nation needed an army to direct public works.

1. *La Technique*

The French tradition began with a military concept of the scientific builder that predated the civilian profession by at least one hundred years. Engineers, literally the soldiers who operated "engines" of siege, were the fort builders and technicians of the seventeenth-century army. Louis XIV saw them as agents of modernization. His finance minister, Jean-Baptiste Colbert, recruited bright engineer-officers to gather statistics, design warships, determine the size and shape of the kingdom, and unify France through a public network of transportation projects. Army engineers paved highways into Paris. They fortified harbors. They completed a link from the Mediter-

Sébastien Le Prestre de Vauban, founder of the French Corps des ingénieurs du génie militaire. Courtesy of the Library of Congress.

ranean to the North Atlantic, Europe's greatest canal. Methodical bureaucrats, they believed in planning and regulation through empirical investigation. They searched for what Condorcet called "the truths of political and moral science," the geometry of government, the laws that removed the caprice from public administration.[11] In seventeenth-century France, as in America a century later, army engineering was a faith in rational bureaucracy where transportation was a public service and science was the compass of state.

What the Americans later hoped to accomplish with a single corps of builders were several branches of engineering in early modern France. The oldest organization was the Corps des ingénieurs du génie militaire. Founded in 1675, it thrived on the legend of its first commandant, Sébastien Le Prestre de Vauban. Vauban formalized warfare through applied mathematics. From the War of Devolution into the War of the Spanish Succession, 1667 to 1703, he perfected the geometry of siege and fortification that accounted for angles of fire. Scale models of his star-shaped defenses became national treasures in a locked wing of the Louvre. His calculations for retaining walls remained in use for generations. His field notes, republished in many formats and plagiarized for apocryphal editions, spelled out the principles of construction later engineers studied at the Parisian highway school, the British academy at Woolwich, and the engineer school at West Point.[12]

Vauban set a standard for army construction in another important respect: he was a master of hydraulic design, and by the year of the great man's death, 1707, his engineers had directed some spectacular waterway projects.

Dunkerque harbor, showing the rounded
piers and jetties designed by Vauban and
completed in the 1680s. Courtesy of the
Library of Congress.

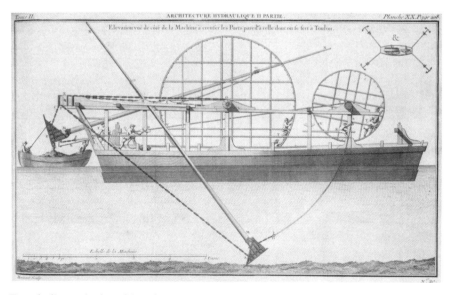

French dipper dredge, driven by
squirrel-cage treadmills, used in the 1730s
on the sand bars blocking Toulon.
Courtesy of the Library of Congress.

One landmark was the massive harbor at Dunkerque in the northeast corner of France. Purchased from the British in 1662, Dunkerque was well situated to command the North Sea access to the English Channel, but shifting alluvial deposits choked off its natural inlet. Vauban cut through these sandy deposits by dredging a deep-water channel between two parallel jetties. At first the builders used long bundles of sticks that were staked into a trench of compacted clay. Gradually they extended the jetties with an elaborate grillwork of rubble-filled, stone-dressed timbers. Rounded piers anchored the structure more than two thousand yards out to sea. A deep navigation lock, added in 1688, kept the channel at high tide. Large enough for forty vessels, Dunkerque became a complex hydraulic system that moated the city and connected the harbor to interior channels.[13]

Dunkerque survived as a showcase for some enduring characteristics of French waterway engineering. First, the harbor was at once commercial and strategic. Its locks supplied a moat that could flood an attacking army. Second, the project reached out to sea with submerged foundations. Vauban learned to work in water with coffer dams, scoop dredges, mechanized pile drivers, and a mortar that bonded with an Italian ingredient—volcanic pozzolana. Third, as a matter of public policy, Dunkerque underscored the rising importance of the militarized nation-state. Louis XIV poured thirty thousand troops into the project. Mustered by cannon at four o'clock in the morning, they marched to the jetties in battle formation. Five hours later another report from the cannon brought in a second shift. With troop labor and the Crown's financial resources, engineers fortified the coastline. Dunkerque, Cherbourg, Calais, Toulon—each port was a naval stronghold, and each commanded interior commerce through public roads and canals.

State financing, military justification, and complex hydraulics remained the outstanding features of French engineering in the king's magnificent network of interior navigation. In 1663 one of Colbert's first acts in office was the establishment of a national survey to study rivers and potential canals. Over the next forty years an aggressive program of land grants and tax concessions added about 180 miles of canals to some 4,800 miles of navigable waterways. Colbert's Canal du Orleans shortened the link between two principal rivers, the Loire and the Seine. The Canal du Centre connected the capital to the Rhône. Parisians used canals the same way American city planners would later use trolley lines: to expand the metropolis, extending its sphere of commerce to undeveloped sites. The French also "canalized" rivers with dams, locks, levees, and towpaths. By 1700 a coastal network of canals and improved rivers, running from Brittany to Flanders, was an interior line of defense against British piracy in the English Channel.[14]

The great triumph of French canal engineering was Europe's first transcontinental waterway, the Languedoc Canal, or the Canal du Midi, a project still used today. Conceived by Pierre-Paul de Riquet de Bonrepos, an engineer, and sold to the king by Colbert, the canal realized an ancient dream: to link the Mediterranean to the North Atlantic by way of Toulouse and the Garonne River, bypassing the sail around Spain through the Strait of Gibraltar. A royal edict launched construction in 1666. Opened in March 1681, six months after the death of Riquet, it ran 149 miles through 114 locks, fifty-five aqueducts, twenty-four culverts for drainage, and fifteen earthen dams. A 540-foot canal tunnel, perhaps the first for a shipping canal ever built, took boats through a ridge at Malpas. Jetties and gates regulated stream flow. A reservoir in the mountains fed an aqueduct system and, near Béziers, an eight-step staircase of locks.[15]

With its novel design, the Languedoc Canal, like the moated port city at Dunkerque, showed the French genius for pipes, dams, and water-driven machines that exploited natural currents. Although the British believed, as Benjamin Franklin later explained, that "locks in rivers are subject to many more accidents than those in still-water canals," the Languedoc connected rivers, the Aude and Garonne, and it moved traffic through the system with floodgates, river towpaths, and other kinds of channel improvements.[16] Following as closely as possible the courses laid out by nature, the canal, an inspiration to Robert Fulton and others, set Americans thinking about their own great waterways through the mountains. Thomas Jefferson spent nine days inspecting the entire length of the project. West Pointer Claudius Crozet, reporting on internal improvements in 1824, thought the Erie Canal suffered by comparison. Where the Languedoc was fed at the summit by a single enormous reservoir, the Erie relied on a crowded system of feeders that made inefficient use of existing streams.[17]

Another example set by the Languedoc project was the active role of the state. The Crown put up almost half the cost of the initial phase of construction, about 7.5 million livres. Colbert, stressing the "public utility" of the project, strived to keep tolls as low as possible.[18] Here began an enduring idea: if government subsidized waterways, the kingdom would prosper and tax revenues would soar. France would pocket a healthy return. Gradually the state learned to protect its investment. Troops filled the labor shortage. Engineer-officers sat on advisory boards. By 1725 about three hundred engineers alternated between civil and military assignments. Now a Corps du génie, they entered the army through competitive examination. Formal reorganization came in 1744. Four years later a statute authorized training for the corps at a new academy, the Ecole royale du génie at Mézières. Government, the patron of waterways, was now the gatekeeper of engineering through technical education.[19]

Although the main mission of Mézières was to train engineers for the army, the academy emphasized peacetime applications of scientific subjects. The two-year curriculum included physics, optics, chemistry, cartography, topography, and architectural design. Soon the school was producing famous names in eighteenth-century science: physicist Charles Augustin Columb, naval architect Jean-Charles de Borda, inventor and balloon aviator Jean-Baptiste Meusnier de la Place, savant of fluid resistance Pierre-Louis Du Buat. In 1775 the Mézières mathematician Charles Bossut applied physics to the study of waterways in a new course on hydrodynamics. Bossut was succeeded by the lowborn physicist and mathematician who became one of France's great teachers and scholars, Gaspard Monge. A prolific researcher, Monge published seminal works on descriptive geometry, the theory of partial differential equations, chemistry, mechanics, and the theory of machines. His pupil Louis Tousard launched the army school at West Point in 1801. Crozet, another Monge student, brought descriptive geometry to the U.S. Military Academy in 1817. The principles of industrial design reached West Point through the American students of Monge's brilliant disciple Jean-Victor Poncelet.[20]

In America the standing of Mézières was matched only by the scholarly reputation of the French academy for bridge and highway engineers, the Ecole des ponts et chaussées. Founded in Paris in 1744, elevated by the fame of its talented superintendent Jean-Rodolphe Perronet, the school developed most of the courses offered at West Point. Lowerclassmen progressed from geometry and trigonometry to mechanics, fluids, conic sections, curvilinear surfaces, stereotomy, and the strength of materials. Seniors studied transportation surveying and architectural design.[21]

Engineer schools in Paris and Mézières also featured hydraulics. More than an academic pursuit, the study of water in France was the search for the scientific principles that would harness rivers, control flooding, irrigate, reclaim swamps, beautify the kingdom, and cleanse its population. Fort engineer Bernard Forest de Belidor wrote the great text on the subject, *Architecture Hydraulique*. First published from 1737 to 1739, and reprinted with corrections in 1813 and 1819, the book was encyclopedic. It covered moats, machine boats, waterwheels, hydraulic cements, lighthouses, harbors, canals, fountains, and hydrodynamics. Although the math was basic and its science was quickly surpassed by more specialized publications, Belidor remained an inspiration. James Monroe returned from Paris with four enormous volumes. Latrobe, distraught, lost his copy at sea.[22]

Frontispiece from Bernard Forest de Belidor's *Architecture Hydraulique* (Paris 1737–1739), showing Parisian fountains and pumps. Courtesy of the Library of Congress.

Belidor's lasting accomplishment was to reinforce the Italian connection between two of the earliest applications of higher mathematics. The first was the study of mechanics as it applied to flood control and river regulation, a science as old as Galileo. The second was the geometrical science of fortifications where water was a weapon of war. Belidor discussed Vauban's experiments with "large water reserves" that, when released at the critical moment, destroyed the enemy's siege. "Since the beginning of the century," Belidor maintained, "the only defense tactics that appear to have made any progress are those that make use of the mobility of water."[23] Engineer cadets explored every aspect of this subject. In the works of Edme Mariotte and Daniel Bernoulli they discovered the science of fluids in motion, hydrodynamics. Their introduction to infinitesimal and differential calculus came through the study of fluid resistance. At Mézières, in the chemistry lab built by Monge, engineers repeated the famous experiment that showed water was H_2O.[24]

Belidor also imported a dark Italian vision of Europe flooded by swamp. This notion evolved from Benedetto Castelli's observation that twisted, sluggish rivers were the most prone to flood. Domenico Guglielmini developed Castelli's ideas in three great works of applied mathematics (1690–1717).[25] Erosion and siltation, in Guglielmini's analysis, were twin agents of destruction that clogged navigation and drainage. The French came to believe that faster rivers were more efficient. Experimenting with levees and dikes, engineers straightened the rivers, closing swamps and unbraiding channels. They also contained high water with straw-covered levees that sloped gently into the river to minimize fluid resistance. Du Buat, endorsing Belidor and Guglielmini, advocated the planting of grasses and trees along rivers to fortify their banks. Ironically, steamer pilots on the flood-prone Ohio and Mississippi later came to the opposite conclusion: trees should be felled and their stumps extracted to prevent logs from blocking the channel.[26]

Back in Paris, and later at the American Philosophical Society in Philadelphia, engineers attempted to verify Guglielmini's theory of rivers through elaborate models. In 1775, Bossut and Condorcet began hundreds of experiments in the Paris laboratory of the École militaire. Tests confirmed that the swifter, straighter channels carried greater volumes of water. Moreover, as width and depth increased, fluid resistance decreased. It therefore required fewer oxen or mules to tow a barge through wider, deeper channels. On the strength of Bossut's recommendation, engineers abandoned their plans for a narrow tunnel on the Picardy Canal. Waterway planners now defended the expense of dredging to at least six feet. Science had altered design.

Pure hydrodynamics remained a mathematical pursuit of little use to the technician. But the outlook of science—its positivism, its need for verification, its indifference to chivalry and noble obligation—was a professional orientation that rebuilt the French civil service, inspiring the savants who transformed public works. At the request of the finance ministry in 1775, Bossut, Condorcet, and hydraulician Jean le Rond d'Alembert formed a waterway advisory board. Bridge and highway engineers became transportation planners in the elite Corps des ingénieurs des ponts et chaussées. Founded in 1720, reorganized in Paris in 1744, the highway corps grew from a drafting office to a bureau of commercial statistics and a clearinghouse for maps, models, and plans. Upgraded again in 1786, it inspected canals, maintained streets, and supervised levee operations. Strict rules of practice (*règles de l'art*) promoted standardization. By the fall of the Bastille in 1789, the Ponts et chaussées had extended the reach of Paris with toll-free highways, river jetties, shipping channels, and about one hundred new miles of subsidized canals.

Critics of the Ponts et chaussées said it promised much more than it built. More planning than action, it focused the kingdom's resources on monumental projects, neglecting routine repairs. Thus, an important stretch of the Loire remained a dangerous torrent. Sandbars crippled Le Havre. Logs clogged the Seine. Dams leaked, and low water in Canal du Centre stranded boats for months at a time. Still, the polish of French engineering dazzled many observers. "France has exhibited some of the most splendid undertakings in inland navigation," said Latrobe's star pupil, Robert Mills. "In whatever direction of France we turn our eyes, we find canals."[27]

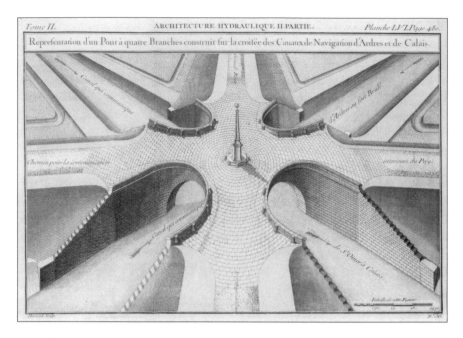

Intersection of the Ardres and Calais
canals, 1739. By 1789 the Crown's bridge
and highway administration was
responsible for about 563 miles of public
canals. Courtesy of the Library of
Congress.

The French Revolution suspended hydraulic construction, but few of the
king's best engineers fell with the Old Regime. Some even profited from the
upheaval. After 1789 the highway program expanded under Antoine de
Chézy and his brilliant successor, Gaspard Riche de Prony. Lazare Carnot,
the engineer-physicist who rebuilt the army during the Jacobin Terror, sat
on the Committee of Public Safety with Prieur de la Côte-d'Or, a fellow
graduate of Mézières. Monge became minister of the navy and then, under
Napoléon, senator for life. He also founded the remarkable school that
spread French engineering and formalized military education in the new
United States: the Ecole central des travaux publics, soon to be called the
Ecole polytechnique.[28]

The Ecole polytechnique quickly established itself as the international
leader in technical education. Founded in Paris in 1794, the school stressed
the mathematics common to civil and military construction, joining these
applications in a single institution. Ties to tradition were strong. Reaching

back to Vauban, the school adopted the military system of weeding out the pampered nobility through a fearsome entrance examination, the *concours*. Students were housed in barracks, organized into companies, and drilled as an infantry corps. About a third of the engineering courses drew examples from the study of water. Prony, the Ponts et chaussées director who taught part-time at the new academy, lectured on hydrodynamics. Professor Joseph-Marie Sganzin featured river works, ports, and canals. Sganzin's lecture notes, assembled in 1804 or 1805, became a fourth-year text at West Point, where it survived in translation until 1837.[29]

Gentleman engineers came to value the polytechnique approach for the two utopian concepts that kept river science alive. First was the scientific belief that rivers, properly understood, would reveal the natural laws that governed all life on Earth. Second, in a more immediate way, the control of water was a political experiment that proved the state could prosper through strict regulation. Prony said "enlightened public administration" would "free the waterways of tolls, which impoverish commerce and choke industry."[30] Uniformity would foster centralization. Government engineers, schooled by the army and employed by the state, would merge the demands of commerce and war in a single, efficient network of transportation projects. Even today, as historian Cecil O. Smith, Jr., has explained, the great feats of French engineering are works of the public sector. State-built and regulated nuclear reactors, the first digitized network of telecommunications, state-operated railroads with the world's fastest trains—these are products of rational planning, each an exhibit of national talent for bureaucracy, science, and standardization.

But the French commitment to government planning was hardly universal. Across the English Channel, another approach to construction was emerging from a private sector that was suspicious of the army, hostile to regimentation, and largely indifferent to the utopian science so cherished by the builder-savants.

2. Brindley and Chalk

"Each Kingdom, each province, each city has its hydraulic needs," wrote Du Buat in his comprehensive treatise, *Principes d'hydraulique* (1779).[31] France needed a disciplined corps of builders to direct public construction, but no counterpart emerged Britain. France needed canals to move Europe's largest army, but preindustrial Britain, a maritime kingdom, relied on a deeply indented coastline with more than a thousand miles of inlets and rivers. In 1750, according to one authority, no marketplace south of Durham was more than fifteen miles from a navigable landing. London became a center of nautical innovation. New kinds of chronometers, compasses, quadrants, and tide gauges fueled the British advance in marine instrumentation. Navigators recorded longitude on the first reliable maps of the North Atlantic. Capt. James Cook chartered the South Pacific. Naval architects, students of French hydraulics, redesigned the heavy warships that reclaimed the English Channel. While the French had brought navigation inland, the British had turned outward to sea.

Before 1750 the British reliance on sea power retarded heavy construction and stifled interior trade. Left behind by the French advance in scientific canalization, the eighteenth-century British simply cut through sharp bends in rivers and drove pilings into crumbling banks. Builders also used a winch and spoonlike shovel to excavate underwater, a technique called "ballasting." Where shoals obstructed navigation, they built movable dams called "staunches," "flash locks," or "weirs." Dutch inspired, these devices had gates to hold back the river until there was water enough to wash traffic downstream. Weirs appeared on the Thames as early as 1585. Arnold Spencer, a seventeenth-century engineer-financier, flashed boats through a hazardous stretch of the Great Ouse River. By the eighteenth-century the British were also experimenting with Italian-style "pound locks" that dammed the river, creating a pool. Although there is no record of pound locks crossing to British North America, weirs and sluices were common. One Yankee variation, the bear-trap dam, used hydrostatic pressure to open wooden gates.[32]

Barton Aqueduct on the Bridgewater Canal, opened to traffic in 1761. Courtesy of the Institution of Civil Engineers.

In Britain and North America the first weir engineers were independent financiers who raised private money and covered expenses through tolls. Aside from a royal charter, a mere endorsement, king and Parliament did little to improve navigation. The royal engineers, reorganized around a military academy in 1741, became the neglected corps of His Majesty's army, poorly trained and understaffed. Its best manuals were French translations. Without support from the army, port cities maintained their defenses while shipping organizations built beacons and wharves. Canals were strictly commercial. When the Irishman Charles Vallancey wrote his treatise on inland navigation in 1763, he looked abroad with envy and deplored improvements at home. "Store of navigable canals and rivers is one of the marks of good policy in a country," Vallancey explained. "In [this] respect Italy, the Netherlands and France, but especially China, abound as much as England and Ireland are defective."[33]

It remained for Francis Egerton, third duke of Bridgewater, to demonstrate the profitability of inland navigation with a canal to his mines at Worsley. Launched in 1759, the first cut was ten and a quarter miles across level terrain. At Barton the canal crossed the Irwell River on three magnificent arches, the duke's "castle in the air."[34] Although larger aqueducts and much longer canals could be found directly across the English Channel in Holland, Flanders, and France, the Bridgewater Canal was different in two respects: it followed a course independent of the rivers, and it minimized the need for canal locks by seeking level ground. "The difference in favour of canal navigation was never more exemplified," wrote John Phillips in his 1792 history of inland navigation. Phillips saw boatmen on the Irwell "labouring like slaves" to draw a barge upstream. Meanwhile on the canal a horse or mule or "two men at most" were pulling "five or six of the duke's barges, linked together."[35] Canals, said Phillips, would counter the degradation of industrialization. Conserving horses, opening new lands, canals would relieve congestion and reduce the high cost of living that crippled the industrial poor.[36]

British canals unleashed an entrepreneurial frenzy fueled by financial risk. At a time when a laborer might make 20 pounds annually, the Bridgewater was a staggering investment, perhaps 220,000 pounds. Yet the canal was a huge success. Extended to the Mersey River in 1767, it opened a water route from the port of Liverpool to Manchester, a rising industrial center. The duke profited from the shipping of lumber, stone, cotton, pottery, grain, and livestock. In Manchester the new canal cut the price of coal exactly in half.

Soon "the fire of speculation was lighted," as the American Robert Fulton put it, and canals became a subject of conversation throughout the British world.[37] A new kind of investor, the canal shareholder, raised nearly

James Brindley, the self-taught builder of
the Bridgewater Canal, painted from a
portrait engraved the year after his death,
1773. Courtesy of the Institution of Civil
Engineers.

10 million pounds in nine years of frantic construction, 1788 to 1796. The
Grand Trunk Canal linked the Rivers Mersey and Trent. Five tunnels and
160 small aqueducts connected the ports of Liverpool, Hull, and Bristol. In
Ireland, 256 miles of excavation brought commerce through Dublin. Americans began on a smaller scale in 1786 with a plan for a 22-mile cut from the
Cooper River to the Santee outside of Charleston, South Carolina. By 1800
there were about 700 miles of English canals. By the 1830s, when the investment shifted to railroads, the British had excavated more than 4,000 miles:
about 3,000 miles of canals in England and Wales, 183 in Scotland, and 848
miles in Ireland.[38]

Although the technology came from Europe, principally Holland and
France, canal engineering in Britain reflected the Whiggish distaste for Tory
centralization. In Britain, unlike France, there were no construction specifications that allowed boats to pass from canal to canal through locks of a uniform size. Because canals were private investments, Parliament made no
attempt to plan an efficient network that linked rivers impounded by locks.
Instead the British simply avoided natural channels. They believed, as Franklin explained, that canals were "quiet and very manageable," whereas rivers
were "ungovernable things."[39] Rivers were destructive giants, either too full
or not full enough. Locks in rivers interfered with mills and they were hard
to finance with tolls. Moreover, experiments by John Smeaton and Thomas
Telford had shown that canals were much more efficient: the horse that
pulled a thirty-ton barge on a river could pull fifty tons on a canal.

When Americans adopted this line of argument, they returned to the pro-canal, antiriver philosophy of the duke of Bridgewater's principal engineer, James Brindley. So much legend surrounds this amazing builder that his work on the Bridgewater project remains in dispute. Some say it was the duke who was directly responsible for the plan of his own canal. Others credit the duke's personal agent, John Gilbert. But it was Brindley who became known in the British world for the traits Americans also cherished. Clever yet common and nearly illiterate, he was "the greatest enthusiast in favour of artificial navigations that ever existed," a student recalled, and it may be that the American concept of the energetic mechanic began in England with this self-made antithesis of the academic engineer.[40]

Brindley entered the profession through a long apprenticeship common to builders and surveyors from the artisan class. According to custom and law, an apprentice served seven years under a master's care before graduating to journeyman's status. He was then free to earn money, travel, marry, and aspire to the status of master in a shop of his own. In Brindley's case the master was a wheelwright and a millwright, the proprietor of a large, inefficient shop where journeymen were content to let new hands embarrass themselves by trial and error. "The training of his inventive faculty and constructive skill was, indeed, a slow but continuous process," his biographer explained, for Brindley could scarcely read "and was thus cut off, to his own great loss, from familiar intercourse with a large class of cultivated minds, living and dead."[41] Not until 1742 at age twenty-six did Brindley establish himself as an all-purpose mechanic in the small market town of Leek near the future canal.[42]

Lack of formal training did not prevent Brindley from inventing machines. He assembled his first engine from memory, patented another, and later devised a system for draining mines. He also invented a canal lock, or, more accurately, he reinvented it, for his solution for lifting boats drew little from Italian designs. Although the duke of Bridgewater, a student of the Languedoc project, was drawn to the science of Europe, Brindley remained an instinctive mechanic. Testifying before Parliament in 1762, Brindley sent out for water and clay to explain his caulking technique in the only way he knew how—by demonstration. When a committee of the House of Commons asked to review the plans of a proposed bridge, Brindley excused himself and returned with a large round cheese. Producing a knife, he cut the cheese in half to represent two semicircular arches. "His resources lay within himself," a brother-in-law explained.[43] Retiring to bed, he preferred to work through a problem in silence, seldom drawing a plan. The eighteenth-century expression "Brindley and chalk would go through the world" referred to this capacity for action that typified the pragmatism of unschooled engineers.

As late as 1771, the *Encyclopaedia Britannica* continued to define "engineer" in the continental sense of a military man with "a perfect knowledge of mathematics."[44] By this definition Brindley was no engineer. The builder of the Eddystone Lighthouse, John Smeaton, was one of the first in Britain to adopt the French *ingénieur* for a civilian occupation, styling himself a "civil engineer." On March 15, 1771, the term appeared at the founding of Smeaton's Society of Civil Engineers. Known as the Smeatonians, they defined "civil engineers" as "a selfcreated set of men," a profession inspired by the "industry" and "steadfastness" of "manufacturing workmen," a calling of "real utility," not of "pompous" or "useless magnificence."[45] The transformation was now complete. A military calling had reappeared in civilian guise—professional yet practical, industrial, and self-consciously British.[46]

The Smeatonians were proof that not all the British builders were unlettered mechanics. Some studied math and surveying in Protestant dissenter academies and nondenominational manufacturing schools. In eighteenth-century Lancashire, for example, an engineer trained as a millwright was "a kind of jack-of-all-trades" with a good understanding of arithmetic, geometry, leveling, and the theory of machines—but these subjects were hardly science.[47] Even the literate Smeaton, whom the Royal Society of London awarded a gold medal for his *Experimental Enquiry Concerning the Natural Powers of Water and Wind to Turn Mills* (1759), used science as a method of promoting or explaining discovery rather than a method of discovery itself. Smeaton had shown that wheels powered by falling water were more efficient than those turned from below by the force of current. Independently, Johann Albrecht Euler and Antoine de Parcieux had reached the same conclusion in Germany and France. But where the Frenchman found a theoretical justification, and the German a mathematical one, Smeaton approached the problem with less abstraction. He simply built a wheel to scale and tested its power.[48]

Smeaton began with models and machines rather than numbers and words. Although he shared a Baconian faith in the linear progress of science, Smeaton, like Brindley, was ready to proceed without it. The chief limitation of that approach was professional isolation. Materials testing, for example, was an impoverished science in Britain. Peter Barlow, a self-taught "master" of mathematics at the Woolwich academy, said works by Euler and others were too abstract, but his own *Essay on the Strength and Stress of Timber* (1817) included errors the French had dispelled a half century before. Barlow was a failure of British mechanics: "another striking example," wrote the historian of science Isaac Todhunter, "of that want of clear thinking, of scientific accuracy, and of knowledge of the work accomplished abroad, which rendered the perusal of the English text-books on practical mechanics published in the first half of this [the nineteenth] century, such a dispiriting, if not hopeless, task to the historian of theory."[49] Yet the popular essay survived for fifty years in six editions.

The few Americans capable of recognizing the inferiority of British texts were those who could read Italian, German, or French. At West Point the study of French remained a prerequisite to the study of construction. In 1817 approximately one of every three books in the academy's collection was in French or translated from the French. From Britain came introductory lessons in arithmetic and general science. From France came "the first really tolerable textbooks on mathematics," said Edward Mansfield, an 1819 graduate. Mansfield claimed that the French methods of teaching science "[are] clear and analytical" whereas "the English treatises are clumsy, being what is called in literature, elliptical, having vacancies in the reasoning to be supplied by the student." At West Point the well-written treatise was a high expression of engineering science. Construction had become "a refined, scientific art."[50]

In Britain, however, it did not necessarily follow that superior books made better engineers. John Rennie, Thomas Telford, George Stephenson, William Fairbairn—these amazing craftsmen were marvels even in France. Rennie rose from a millwright to the engineer known for some of Europe's most massive projects: London and Waterloo bridges, Rochdale and Lancaster canals, towering Bell Rock Lighthouse, and the mile-long breakwater that walled off a harbor in Plymouth Sound. Telford, once a mason, won equal acclaim. His Caledonian Canal opened a passage through Scotland, and in 1805 he finished a Roman-style nineteen-arch aqueduct that took boats through an iron trough. Although Telford's notes on mills and reservoirs earned some notice from scholars, he was generally bored by theory. A clue to his orientation survives in the papers of his professional organization, the Institution for Civil Engineers. Begun in a London tavern in 1818, the society collected "practical" descriptions of "works really executed." These builders wanted "facts," not "learned discussions." "We admit only what has really been performed," Telford insisted, not "theoretical projects."[51] In Telford's London, where there were no competitive examinations for engineers, experience outweighed academic speculation.

Soon the Institution for Civil Engineers had built a respectable technical library that closed some of the gap between French and British builders, but across the Atlantic the contrast was sharp. Americans, to paraphrase Latrobe, were tinkers or thinkers but rarely both. While tinkers attempted to simplify science in British-style lyceums, the poet Joel Barlow, an American adventurer in Paris, promoted a highbrow Polysophic Society to disseminate

French ideas.[52] And while thinkers retained the Parisian bias for scientific planning, tinkers guided construction with a less visible hand. In America, a divided republic, waterway engineers were evolving from two professions that competed but seldom converged.

3. The British Yoke, the Gallic Chain

Nowhere were the professional alternatives more distinct than in the technical forces under General Washington's command in the ragged army of revolution. It had taken Washington just seven days as commander in chief to bemoan "the want of engineers" to supervise fortifications and the placement of artillery for the Continental Army. Congress had granted the first engineer commission to Col. Richard Gridley, an artillerist with the Massachusetts militia. At age sixty-five the colonel was past his prime. Washington preferred a younger man, the millwright Rufus Putnam. "Altho' he is not a man of Scientific knowledge," Putnam, a lieutenant colonel, was "indefatigable in business" and Washington claimed he had "more practicale knowledge in the art of engineering" than any man in the American camp.[53]

But the millwright knew little about fortifications. As the Americans took the high ground above Boston in early March 1776, Putnam searched for a

Opposite page:
John Rennie's mile-long Plymouth Breakwater, begun in 1812 and completed by the engineer's son in 1847. Courtesy of the National Maritime Museum, London.

Thomas Telford with a background view of the Pont-y-Cysylltau Aqueduct on the Ellesmere Canal, one of his great works. Courtesy of the Institution of Civil Engineers.

way to fortify Dorchester Heights. Typically a frontier army would crouch behind a fence of logs anchored by wooden stakes. The hills, however, were frozen and Putnam could not drive the stakes. Europeans solved this problem by building a temporary fence with tightly bound sticks ("fascines") stacked into wooden frames ("chandeliers"). One night Putnam chanced upon these curious words in an English translation of a French field manual, one of Washington's books. The millwright later recalled "those singuler circumstances which I call providence" that held the solution to Dorchester Heights:

> I took my book from the Chest, and looking over the contents I found the word, Chandilears. What is that thought I. It is somthing I never heard of before, but no sooner did I turn the page where it was described with its use but I was ready to report a plan for makeing a lodgement on Dorchester Neck (infidels may laugh if they please).[54]

Through the night of March 4 the Americans moved in behind Putnam's first chandeliers. Heavy rains delayed the battle. When the sky cleared on March 7, Washington's position seemed strong behind fascine fortifications. Ten days later the British sailed for Halifax, evacuating Boston.

Putnam's fascines and chandeliers point to a crack in a modern assumption about technology transfer. Because the transit seems to rely on direct,

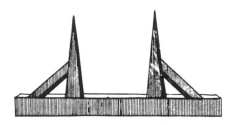

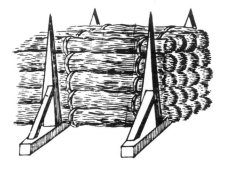

Chandeliers, from L'Abbé Deidier, *Le Parfait Ingénieur Francais,* Paris 1757. Courtesy of the Office of History, U.S. Army Corps of Engineers.

word-of-mouth, builder-to-builder contact, historians tend to minimize the importance of written words.[55] But writing can sometimes spark technological adaptation. What Putnam called providence was merely a book.

It was also true that Putnam consulted a book as a last resort. In 1776 there were no experienced fortification engineers in America save those in the enemy's camp. Even millwrights were in short supply. In 1776 the Philadelphia Council of Safety entrusted the defense of Red-Bank on the Delaware to a novice colonel named John Bull. Bull's defenses were strong but poorly placed. French officers damned these works with faint praise, commending the Americans for "natural good sense unenlightened by theory."[56] The forts at Red-Bank were frontier defenses inadequate for America's capital city, and Putnam, at least, recognized the need for a more professional class of builders. Congress had already appealed to Louis XVI. In April 1776, Silas Deane sailed for Paris with orders to beg reinforcements, purchase munitions, and recruit good engineers.[57]

In four years of heavy fighting from 1777 to 1781, the French came to America from Europe and the Caribbean in a great wave: about seven hundred army officers and twelve thousand soldiers joined more than thirty-two thousand French sailors from fleets off the American coast. Some were Frenchmen of "low character" who, said a British agent, "behave with great impudence." Some were "gallant officers" with rooms in places like New-

Louis Lebébue Duportail by Charles Willson Peale. Born to a noble family and educated at the military academy at Mézières, Duportail became Washington's engineer commandant on July 22, 1777. Courtesy of the Independence National Historical Park Collection.

port where they dined with the rich.[58] At least sixteen called themselves engineers. Most had studied at Mézières. Several were able to secure commissions before booking passage, while others, taking their chances, arrived as informal aides. Artist and future engineer Pierre L'Enfant volunteered at his own expense, hoping for promotion. As young aristocrats and gentlemen soldiers, many insisted on horses, carriages, servants, and rank befitting their station. Engineer and artillery expert Phillippe Charles Tronson du Coudray, Washington's choice for engineer commandant, would accept nothing less than a major general's commission. When his horse leaped off a floating bridge, Coudray died in the Schuylkill River, and his commission passed to Louis Le Bègue de Presle Duportail.[59]

Duportail, a proud man, had attended Mézières about the time Monge was a teaching assistant. Recruited by Deane, he became a colonel and then, in 1779, commandant of the reorganized Corps of Engineers and Sappers and Miners. The following year he fell captive to the British, but Washington won his release through a prisoner exchange. Although loyal to Washington, Duportail spat contempt at the "effeminate" American army. Pampered and unpatriotic, the Americans, said Duportail, were lowborn mechanics and tightfisted tradesmen who spent their time smoking and drinking. American troops seemed to prefer "the yoke of the English to a liberty bought at the expense of life's amenities."[60] As for clever craftsmen like Putnam, they were

Allied French and American forces
storm British position at Yorktown
in this lithograph by an unknown
French artist, circa 1781. Courtesy
of the Library of Congress.

Washington's plan of attack
at Yorktown, drawn by French
engineer Jean Baptiste de Gouvion,
1781. Courtesy of the Office of
History, U. S. Army Corps of
Engineers.

"engineers only by name." Duportail said Washington needed a builder of vision, "a Vauban," to unite artillerists and engineers and organize "an academy" for "men of theory and knowledge."[61] This advice to the army launched two important ideas. One reached fruition in the 1794 revival of the fortification program under the U.S. Corps of Artillerists and Engineers. The other bore fruit in 1802 at the engineer school at West Point.

Duportail, by his reports and personal example, underscored the essential distinction between self-taught craftsmen and academy-trained engineers. Although contentious and painfully conscious of status, the French commandant was also a fine tactician, even, Washington insisted, a "military genius."[62] At Philadelphia and Valley Forge he revitalized crude defenses, using the science of Mézières to calculate angles of fire. Joining Washington and Lafayette at Yorktown, Virginia, he plotted the siege of Cornwallis that forced an end to the war. Yorktown was the brilliant success of America's French alliance. With nine thousand troops on the ground and about twenty thousand French sailors off the coast of Virginia, the French outnumbered American soldiers three to one.[63]

Duportail wrote a final report for Congress and returned in 1783 to the rank of lieutenant colonel in the ill-fated army of Louis XVI. Eight years later he was a high-ranking member of Robespierre's government and, briefly, minister of war. In America, meanwhile, the cause of the French engineers fell to junior officers like Jean Baptiste de Gouvion and Pierre Charles L'Enfant. "It is not very difficult to form an officer of foot, or horse," said Gouvion, a veteran of Yorktown, "it does not require a long Space of time." But the engineer or artillerist needed "a well-established military academy" where "his Knowledge of mathematics must be carry'd to perfection."[64] L'Enfant went much further. In his 1784 plan for a peacetime army, the Frenchman proposed a department of public construction, an American Ponts et chaussées. Academically trained, the corps would study arithmetic, geometry, mechanics, architecture, chemistry, and physics. It would also investigate "water works" and "the means of raising or changing the course of water." Its jurisdiction would embrace "every work which the States Should considere as necessary" including "an atlas of the whole Continent" and "all military and civil building"—forts, bridges, roads, and canals.[65] Breaking the nation's dependence on foreign officers, the peacetime corps would keep engineers employed and ready for military assignment in the event of sudden attack. Here, in a letter to Congress, was the enduring connection between preparedness and transportation projects, the modern rationale for Corps supervision of federal public works.[66]

Forty years would pass before L'Enfant's vision took shape in the army's work on roads and canals. By 1784 the soldier-engineers had already disbanded. Nationalism of any sort, even a national army, was out of season. Congress reflected the political mood by placing internal improvements outside the federal jurisdiction. Washington made a statement by retiring to Mount Vernon, and there he delighted his guests with plans for a canal across the Alleghenies to the Ohio River, a project for private enterprise and civilian builders. Militias would keep the peace. Local contractors would repair bridges and roads. As for the French engineers, most sailed for Europe or the French Caribbean, although a few returned in exile from the Jacobin Terror. Duportail bought a farm near Philadelphia. L'Enfant drew plans for the new federal city on the Potomac. Others, called back by Washington in 1794, directed a rush program of coastal fortification.

The vacuum left by the French made room for a disparate array of aspiring builders. There were carpenters, masons, instrument-makers, self-proclaimed architects, and gentlemen surveyors and astronomers from the American Philosophical Society. There were also promoters and inventors like Robert Fulton and Oliver Evans, but most underestimated the complexity of hydraulic construction. On the Ohio frontier, a visitor later reported, "the first-comer in a bar-room will explain to you, over his glass of whiskey, how to feed the summit level and how to construct a lock." Canaling was becoming mere excavation, "an affair of the common people."[67] Yankee builders were shrugging off methodical planning. Wood was replacing stone. Haste was eroding the polished workmanship of Europe. David Stevenson, a Scottish engineer, crossed the ocean to study these primitive projects. Landing in New York in 1837, he set foot on a muddy plank, quite taken aback. There were no masonry piers, no iron supports, no covered verandas, but gradually the Scotsman began to appreciate the American situation. Timber was plentiful. It rotted but was cheap to repair and replace. Why build with stone only to build again when the dockyard moved or expanded? In America it made sense to keep pace with growth by tacking on additional planks.[68]

Wood narrowed distinctions between native factions of construction experts. Log piers, plank canal locks, wooden aqueducts, timber dams—all marked the rise of the frontier builder during America's wooden age. Some Americans liked to equate these innovations with political freedom and technological independence: the "young and rising states," wrote the poet Jezaniah Summer in 1798, would unharness the "British yoke" as they severed the "Gallic chain."[69] But water seemed to cool the native genius. In 1816, for example, the Erie Canal commissioners searched in vain for a qualified builder. Desperate for direction, they inspected the shallow cut between the

Pierre Charles L'Enfant's famous plan for the city of Washington, 1791. Courtesy of the Library of Congress.

Below:
Canal locks at Manayunk, Pennsylvania, pictured here in an 1830 woodcut, show the preference for timber construction during America's wooden age. Courtesy of the Library Company of Philadelphia.

Merrimack and Charles rivers, the Middlesex Canal. Its first engineer had once been a cabinetmaker. He was baffled by hydraulics, and his canal bed leaked. So did his locks. Yet the Middlesex was proclaimed "the best artificial navigation in the United States," and the Erie commissioners cited the project as proof that "construction" was "an art within the complete attainment of ordinary capacities."[70] Clearly it was not.

Before West Pointers entered the field, foreign counsel was chiefly British, and it entered America through immigrant craftsmen and surveyors who found work as engineers. One was John Christian Senf, a British colonel who was born a Swede (though Latrobe thought he was German). In 1786 Senf took charge of South Carolina's Santee Canal. Senf underestimated time and expense, and the project could not turn a profit. Another "engineer" called himself James Brindley and said he was kin to the great Brindley in England. From 1786 to 1797, his American failures included the Susquehanna and Conewago canals. George Washington, elected president of the Potomac Canal in 1785, held out for "a proper person from Europe," possibly a Frenchman, but the investors wanted a builder with a good command of English.[71] The search turned up Senf, the lesser Brindley, steamboat inventor James Rumsey, miller Thomas Gilpin, artisan Christopher Myers, and others but no engineer. Meanwhile, on the Mohawk River, the Western and Northern Inland Lock and Navigation Companies had hired and fired two British millwrights, Archibald Nisbet and James McCotter. Latrobe resented these kind of pretenders, calling them "villainous quacks."[72]

Then, in January 1793, the Englishman William Weston arrived in Philadelphia. Perhaps no master by the standards of Europe but certainly no "quack," Weston had worked on the Oxford Canal and rose as a bridge engineer under one of Latrobe's famous mentors, William Jessop. Weston, apparently recommended by Jessop, came to America at the invitation of several Philadelphia companies competing for the western trade. One would join the Schuylkill and Susquehanna rivers from Reading to Middletown, Pennsylvania. Another projected a sixteen-mile cut along the Schuylkill from the Delaware to Norristown. A third wanted a short canal around the falls of the Susquehanna near the mouth of Swartara Creek. There was also a turnpike planned. Weston moved from project to project with disregard for the caution of science—surveying on horseback, producing plans and estimates on site. He tore out half-finished locks, replacing wood with stone. He loaned out his English shovel. He taught Americans to level with the proper tools. He explained the wheelbarrow and followed British practice by organizing companies of workers under overseers. He also imported the British method of caulking or "puddling" a canal with layered applications of clay.[73]

By 1801, when Weston returned with his wife to England, he had assisted and advised at least a dozen ambitious projects: the Middlesex, Potomac, Conewago, and James canals; also a short canal to Rome on the Mohawk River; the Lancaster turnpike; and Philadelphia's Market Street Bridge. He left some of this work to competent students like Loammi Baldwin, Sr., Robert Brooke, and Benjamin Wright. His influence on the army was felt through Isaac Roberdeau, a British-trained Weston assistant who became chief of the future U.S. Topographical Bureau. Other water projects fell to consulting engineers from Britain like George Duncan and Thomas Oakes. Meanwhile the Americans Loammi Baldwin, Jr., and Canvass White went to England to study canals. These builders transmitted technology in the most immediate way. In 1816 the Erie Canal commissioners in the Mohawk Valley were still consulting Weston's survey.[74]

With British engineering came folkways and inclinations that were foreign to French hydraulics. Weston, like Brindley before him, was a strong advocate of level, dead-water, dry-land excavation. Reporting to the Schuylkill and Susquehanna Navigation Company in 1794, the Englishman scuttled plans for locks and dams in the Tulpehocken River. The "unerring test of experience," Weston explained, had shown the folly of slack-water projects, especially in America where "the danger" of river canalization was "much greater than in England, as the floods are more violent, and accompanied by ice in greater quantities."[75] As builders began to abandon slack-water projects—on the Potomac, Mohawk, and James, and wherever a canal bridged or turned deliberately away from the natural channel—Americans cited the British gospel in two familiar parts: first, minimize lockage; second, use rivers to feed canals.[76]

Even when builders held out for channel improvements, there were few publications with plans and, until 1837, no American books with sections on dams and jetties. Weston was sure that "practical examples are more conclusive than theoretical arguments."[77] He gave America tools and site-specific reports rather than journals and textbooks. Before the 1820s, when a few French works began to circulate in translation, a builder might read something useful in a small shelf of British mechanical periodicals in the Philadelphia Library. He might find something in Smeaton's essays or Fulton's brilliant book on canals, but otherwise the English-language literature on civil engineering was full of promotional tracts. Erie engineer John B. Jervis looked under "bridges," "canals," and "carpentry" in his uncle's treasured copy of *Edinburgh Encyclopedia,* a rare source. In 1819 the *North American Review* lamented that engineering books and instruments were expensive and scarce: "if any person seeks for them in the shops or book-stores in the

United States, he will be disappointed."[78] West Point was in its third decade before a rival academy published a simplified text for civil engineers.[79]

Inadequate books and technical schools deepened the basic distinction between an engineer's craft and his science. Craftsmanship, more ancient than science, relied on personal and sustained builder-to-builder contact. It interacted with science but thrived with or without it. Craftsmanship was the Brindley-like genius for tools and machines and the talent for brute excavation that gave level-headed builders their preference for level canals. But not every American builder shared that orientation. Where *la technique* was strong—at West Point, in Franklin's philosophical society, among officers and surveyors in Latrobe and Gallatin's circle—engineers still plotted on the grandest scale. Their hope was in hydrodynamics. If science could explain stream behavior, then surely, these planners maintained, canals could link rather than avoid river systems, jetties and dams could improve navigation, and a corps of methodical builders could unlock the natural channels that were America's great highways of trade.

4. Philosophers and Working Mechanics—an Enduring Dispute

"An engineer," said Henry Robinson Palmer, speaking in London at the 1818 founding of the Institution for Civil Engineers, "is a mediator between the philosopher and working mechanic, and like an interpreter between two foreigners, must understand the language of both."[80] For builders in distant America, that image was apt. Like interpreters, the Americans transmitted foreign ideas, but rival orientations drew builders to competitive camps. While West Pointers dressed the part of the polytechnic scholar—white pants, high collars, brass buttons, plumes in their hats—civilians, with few exceptions, preferred the frock-coated tradition of Anglo-American mechanics and company engineers. Such diversity foiled scientific attempts to elevate technical standards. Waterway engineering was still a fluid profession where different groups advanced dissimilar technologies independently and contemporaneously.

The fluid and fragmented world of the early surveyor and builder reflected an ambivalent link to the original source of America's craft tradition, Britain itself. Although many Americans despised the British connection, the two nations shared a language, a legal system, a work ethic, a talent for mechanization, and other ties that made Yankee construction receptive to British ideas. Thus, historian Anthony F. C. Wallace has referred to a two-part "technological ethos" that bound the United States to its cultural parent: first, a British resistance to scientific abstraction; second, a porous social structure that rewarded lowborn inventors, allowing artisans to work efficiently with gentleman entrepreneurs.[81] No similar technological ethos made headway in an American army where engineering blurred into science, where an officer's commission was often a political appointment and highborn builders like Duportail held artisans in pained contempt.[82]

Cultural paranoia widened the technological rift. At one level, perhaps, the fissure through engineering was a military fear of the British, a suspicion hardened by war. Elsewhere the hostility was directed at France. Although Americans, many of them, remained grateful to Lafayette, most had reacted with horror when the promise of the French Revolution soured during the Terror. Federalists, the party of Washington and Hamilton, compared the French Revolution to a deadly disease from the French Caribbean. In August 1793, for example, when yellow fever hit Philadelphia, the Federalist press turned on the French. Had French refugees poisoned the wells? Had they joined the Republicans in some sinister plot to close Philadelphia's trade? The hysteria peaked in 1798 during the undeclared war with France. As Napoléon disrupted American shipping, two excellent engineers—Jean Foncin and Chevalier de Rochefontaine—were roughly expelled from the army. A third Frenchman, the loyal Tousard, was passed over for promotion. Meanwhile the Adams administration was calling Jefferson's people "the French party," a cabal of intrigue and sedition. Jefferson traveled under close surveillance. His ally Aaron Burr chided the opposition: if the Federalists were afraid of French engineers, they should watch for a royalist coup—the émigrés at American forts had once served the French Crown.[83]

The vast historical literature on these formative years of the army has neglected the cultural and technological conflict left by the Federalist Era. Captivated by the idea that a well-trained engineering corps could serve the needs of commerce as it readied the nation for war, historians often miss the civilian distaste for the French aristocratic tradition. Forest Hill's book on the Corps called West Point engineering "the technology of the nation."[84] Charles F. O'Connell, Jr., agreed that the Corps was a catalyst to industry and a management model for the nation's first big corporations. And while the historian Daniel Calhoun was aware of essential distinctions between scientific officers and field-trained canal builders, he denied that there ever developed "an overt, general competition" between these professional groups.[85] Thus, in scholarly writing, the Corps is usually found at the forefront of a cohesive engineering profession—procapitalist, pro-British, divorced from the bitter wrangling of national politics, and receptive to innovations embraced by American builders at large.[86]

In Jeffersonian times, however, the West Pointers were mostly outside British tradition. Professional division was tense, often overt, and the army's contempt for civilian construction was as old as the first contact between French and American builders. In 1793, for example, the émigré engineer Pierre Pharoux noted the crude "incompetence" of New York canals. Elsewhere Yankee construction was "dreary," "a mass of loose irregular stones."[87]

Likewise, in civilian life, a nativist fear of the stranger isolated the émigré French. In 1813, Latrobe referred to "a violent prejudice among the Federalists against everything French," especially against French engineers like his good friend Godefroy, an underemployed Parisian.[88] Three years later, while the state of Virginia was searching for a superintendent of public works, Brig. Gen. Joseph G. Swift said there were no qualified Americans, and he suggested a letter to Prony in France. But that year the army hired one of Prony's students and Swift, in a rage, resigned. Such episodes launched a wider attack on army programs—a scorn for regimentation, a resistance to West Point attempts to modernize arms production, a history of friction between the Corps and civilian technicians in the U.S. Lighthouse Establishment and U.S. Coast Survey, a fear of militarism that continues even today. "West Point traits are almost diametrically opposite to those required in large-scale civil engineering," wrote engineer Arthur Morgan in *Dams and Other Disasters* (1971), a bitter indictment of army tradition.[89] Engineers, Morgan acknowledged, imported this conflict from Europe. Professional division endured.[90]

At the same time there were crosscurrents in Jeffersonian America that blurred some of the lines between the two schools of builders. Mechanical institutes, less status conscious than workingmen clubs in England, were beginning to offer scientific education to a literate class of inventors. Another modest departure from British tradition were the semipublic "mixed enterprises" that combined privately held stock with government loans. But how could a canal company solicit federal aid without bringing in federal experts? And how could the promoters skirt the constitutional restriction against aid to corporations?

One partial solution was half thought out by L'Enfant, an idea both British and French. Army topographers could draw an atlas of the continent. Mapping navigation, they could stop just short of construction, planning without building, promoting commerce and educating the nation through the science in their surveys.

2

Mapping Water, Marking Land

Henry A. S. Dearborn of Massachusetts, attorney and canal investor, was at a loss to explain why Congress had squandered the talent of its best water surveyors, the U.S. topographical engineers. "Hitherto," wrote Dearborn in a book on western commerce (1839), "the [topographical] officers have been chiefly engaged in reconnoisances and surveys, and have not been favored by the opportunities of carrying into effect the various extensive plans for pub lic works." But given the chance to build, these officers would complete the "golden chain" of navigation that bound the West to New England. Topographers, said Dearborn, were "exact in preliminary investigations, ingenious in theoretical modes of construction, perfect in the details of plans, and accurate in estimates."[1] They had joined the Corps of Engineers in the army campaign for French-style transport planning, yet men like Dearborn could see that the topographical engineers were also a link to the British example of the military geographer, a wilderness scout. Heirs to this hybrid tradition, the topographers grafted surveying to science, transforming the study of rivers with a frontier mix of ideas.

Although good topographical officers had been highly prized by the British, the new United States in its first four decades did little with army surveyors. Only a few skilled draftsmen and scouts had been found for the U.S. Geographer's Department that mapped for Washington's army during the Revolution. Disbanded in 1783, some surveyors returned to government service as fort planners, lighthouse experts, and boundary commissioners. When the military reorganized in 1813, Congress created the "Topographical Section of the War Office," a force of eight officers and eight assistants. Topographers would "make plans of all military positions . . . accompany all reconnoitering parties . . . make sketches of their routes . . . keep a journal . . . [and] exhibit the relative positions of the contending armies on the fields of battle."[2] The army filled ten of the sixteen topographical positions during the War of 1812. As the War Department reorganized after the Treaty of Ghent, Congress, in 1816, authorized three topographical engineers and two assistants for each of the army's two divisions—ten officers in all. In 1818 the War Department placed a major in charge of a new "topographical bureau." For the next thirteen years the bureau was attached to the parent Corps of Engineers in a single "engineer department" under one "chief engineer."[3]

The merger of the bureau and Corps was a sign of their common purpose. Both agencies planned internal improvements and both, increasingly, attracted top graduates of the U.S. Military Academy. Although only two of the original ten topographers were West Point graduates, thirty-eight of forty-five had West Point diplomas by the end of the Mexican War. So close were the bureau and Corps that historians have treated them almost as one. W. Turrentine Jackson's *Wagon Roads West* (1952) found "no attempt" by the War Department "to specify the duties of the Corps of Topographical Engineers as distinct from those of the Corps of Engineers."[4] Another writer dismissed "fine points of distinction" between two kinds of builders who were both, after all, "performing duties that might be called 'civil engineering.'"[5]

Building and surveying did overlap, but the more the bureau was confused with the Corps, the harder both worked to preserve autonomous status. Maj. Joseph G. Totten, a Corps fort engineer, feared topographical surveying for what it might become: a civil-works program, a diversion from the more urgent matter of national defense. In 1824 the major resisted a plan to strengthen the Corps with half-civilian brigades of topographical engineers. If, Totten insisted, the surveyors joined the builders, "a constrained & unnatural union" would cripple engineering, splitting the Corps in two.[6] Soon the topographers were deeply involved in programs essential to Congress but distant from the combat orientation of the army's high command. Even after 1838, when the U.S. Topographical Bureau became the U.S. Corps of Topographical Engineers, the chief of the enlarged organization still pined for recognition. "Ours is . . . a corps of no exclusive privilege," wrote Col. John J. Abert in an 1840 letter to Van Buren's War Department. "We have always labored at serious disadvantage having once been subordinate to and under the command of the Corps of Engineers."[7]

Abert's complaint reflected the fact that the *crème de la crème* of West Point cadets preferred to build fortifications. Surveying was second choice. West Pointer John Tidball, a graduating senior in 1848, recalled "the utmost reverence" for this hierarchy, "a kind of fixture in our minds that the [Corps of] engineers were a species of Gods, next to which came the 'topogs'—only a grade below the first, but still a grade—they were demigods."[8] Steeped in a splintered tradition and tainted by civil works, the topographers were more than chain surveyors but something less than elite. They became, as Totten predicted, the army's civil surveyors, a flexible force of half-breed soldier-civilians that strived for professional status as a bridge between technical worlds.

1. Compass and Chain

"Surveying," wrote Andrew Ellicott in 1796, "has been attended with so much inaccuracy, and although the work already executed may not admit of a complete remedy, that which remains to be done may be worthy of attention."[9] Ellicott, one of America's most scientific surveyors, was a U.S. boundary commissioner, the man sent west by President John Adams to map the frontier of New Spain. An accomplished astronomer and mathematician, Ellicott longed for the day when the art of measuring land became a methodical science. Ellicott hoped that the blend of surveying and science would disseminate new kinds of equipment—slide rules, tide calculators, moondials, instruments that measured air pressure and water vapor, tools that tracked Venus and read solar time. Science would bring globes and relief maps, textbooks and technical journals. And science, fostering innovation, would fuel a zeal for public improvement that bound the Republic with roads and grounded the mapping of rivers in the study of currents and tides.

Ellicott wrote about the promise of science in an age when most common surveyors still relied on the compass and wrought-iron chain. If a surveyor also had a theodolite, he could measure the vertical distance between two points—the foot of a hill, say, and its summit. The incline became the hypotenuse of a right triangle, and altitude could be determined by consulting tables for the ratio of slope to height. Most frontiersmen, however, preferred traverse surveying, a method of figuring angles and distance by compass bearings alone. One man who promoted this method was the British author John Love. A plantation surveyor, Love toured Jamaica and Chesapeake Bay before returning to London in the 1680s to publish a popular handbook. The book, aimed at American readers, included "plain and practical rules" for surveying in frontier conditions. It also had sections on rivers and harbors. "Measure first the sea-coast on both sides of the river's mouth," Love instructed; "then going out in a boat to such sands or rocks as make the entrance difficult, at every considerable bend or [at] the sands, take with a sea-compass the bearings . . . of two known marks upon the shore."[10] The method, crude but effective, allowed men to approximate the contours of coastline, but surveyors still needed scientific equipment to mark boundaries across rugged backcountry terrain.[11]

Falls, rapids, and swamps challenged the long-distance surveyor. One early swamp expedition inspired a comic memoir, William Byrd's *History of a Dividing Line Betwixt Virginia and North Carolina Run in the Year of Our Lord 1728*. A vivid look at the coarse frontier through the eyes of Byrd, a genteel planter, the *History* follows a boundary commission adrift in the Chesapeake wild. Byrd found savage men and savage beasts and, most forbidding of all, giant bogs where his companions "frequently sank up to their middle without the least complaint." The most treacherous ground was the wet depression between Chesapeake Bay and Albemarle Sound, the Great Dismal Swamp. "When we got into the Dismal itself," Byrd reported, "we found the reeds grew there much taller and closer and, to mend the matter, were so interlaced with bamboo briers that there was no scuffling through them without the help of pioneers." The ground was "moist and trembling under our feet."[12] Equipment disappeared. While the guides cleared a path with their hatchets, Byrd retreated to higher ground. Nine days passed without a word from the surveyors until at last they emerged in a frightful condition. In all, the party had crossed about fifteen miles of swamp in fifteen sweltering days.

Washington as a Surveyor, an engraving by Octavius Carr Durley in Washington Irving's *Life of George Washington* (New York 1857). Courtesy of the Library of Congress.

In Byrd's account of the expedition, literary scholars point out, the term "dividing line" is rich with mulitiple meanings. A path to Eden, a cultural demarkation between tidewater society and the coarse frontier, the surveyor's line was also the status line between noble and common, gentle and base. Gentlemen scouted and made calculations. Commoners carried the chain. In the Chesapeake region there were high-born surveyors like Byrd, Robert Beverly, George Washington, and the Reverend James Madison, first Episcopal bishop of Virginia, who found time in his ecclesiastical career to map the boundaries of the commonwealth. There also were lowly mechanics. Latrobe called them "good, plain sensible men," artisans who sometimes worked their way into the profession as axmen and rodmen on road and canal surveys.[13]

To the north the profession made room for gifted machinists who learned surveying as instrument makers. David Rittenhouse followed this course, as did the Englishman John Smeaton. Benjamin Franklin's American Philosophical Society became a clearinghouse for these practical astronomers who fixed latitude, longitude, and local time. Some surveyors learned to read stars in the classroom. In 1765 the New Yorker Thomas Carroll advertised a "theory" course that taught "two Universal Methods to determine the Area of right lined Figures, and some useful Observations on the whole; . . . Navigation; . . . keeping a Sea-Journal . . . Astronomy; [and] Sir Isaac Newton's Laws of Motion."[14] In Philadelphia on the eve of the Revolution, a surveyor could study with Robert Patterson, the Irish-born mathematician who later taught the practical side of astronomy at the University of Pennsylvania. Christopher Colles, another Irish surveyor, came to America hoping to work as a hydraulic engineer. Struggling as a science tutor in Philadelphia and New York, he taught astronomy door-to-door.[15]

Military surveyors matched the diversity of the profession at large: there were gentlemen and commoners, astronomers, engineers, and those who just chained the land. During peacetime the best of the British officers drew maps that left nothing to the uncertainty of the compass, surveys that accounted for magnetic variation, atmospheric refraction, and the curvature of the earth. Engineers Samuel Holland and Joseph F. W. Des Barres set new standards for American cartography as foreign-born officers in Britain's Sixtieth Regiment, the Royal American Regiment of Foot. Trained in Europe and London, Holland and Des Barres planned forts and led well-equipped survey parties along the northeastern seaboard and the St. Lawrence River. At Louisbourg, Nova Scotia, Holland and Des Barres taught the use of a plane table to one of the empire's great explorers, Capt. James Cook. Together they used instruments as rare as astronomical clocks and a twenty-four-inch

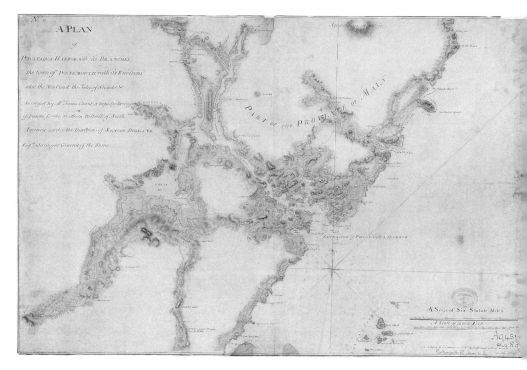

Above:
Piscataqua Harbour at Portsmouth, New Hampshire, surveyed by British engineer Samuel Holland, 1773. Courtesy of the National Maritime Museum, London.

British hydrographer Joseph F. W. Des Barres took precise soundings of the St. Lawrence River for his "Plan of Quebec and Environs," 1759. Courtesy of the National Maritime Museum, London.

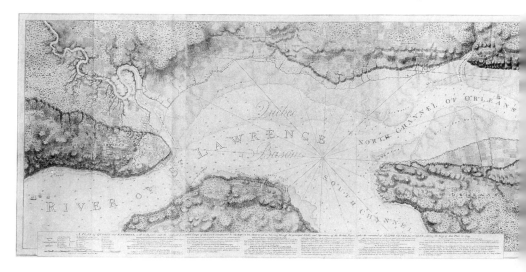

reflecting telescope. They experimented with barometers and portable sextants. They studied soil and vegetation, taking samples. By 1775, when the American war interrupted exploration, Holland, Des Barres, and others were fusing army surveying with an interest in natural science well beyond strategic concerns.[16]

The more common product of frontier reconnaissance was a map of an army's march, a sketch of its defenses, and a chart of water hazards in special detail. British commander Thomas Gage ordered his surveyors to study America's rivers, "remarking breadth and depth and the nature of their banks, the fords, if any, and the nature of their bottoms."[17] From 1763 to 1775, years of intense exploration, British surveyors charted almost all the navigable reaches of the Atlantic tidewater rivers. Turning inland, they followed the St. Lawrence to the lakes, the length of the Ohio, and the course of the Mississippi from New France to New Spain. Sometimes the major channels were enlarged from a scale of one to four inches per mile, a practice adopted by Anglo-American surveyors and continued with some variation by the future topographical engineers. Holland, for one, worked the northeastern seaboard, occasionally tracing the rivers but seldom looking beyond. In this way the British neglected the heartland. Their attention to waterfront was consistent with the military assumption that few colonists would brave the dark interior. A British official summed up the army's position in 1764: "Attempting to survey more than the harbours, the coast, the navigable rivers, and the open roads will be putting the Crown to an expense for little good purpose."[18]

Laconic and focused on the perils of navigation, the British accounts of interior rivers were typically a genre apart from the fanciful voyage literature of eighteenth-century Europe. Philip Pittman, a British lieutenant, mapped the lower Mississippi and assessed its river defenses in a journal that ignored hearsay accounts of a northwest passage. His chief concern was the strength of the Spanish garrison at New Orleans—its troops, guns, and fortifications. This kind of report gave an army a huge advantage. In July 1780, while General Washington was planning a raid into Connecticut, he searched for a chart of the seaboard: "It is a matter of great importance for me to be acquainted with our several harbours, their depth of water with the leading to them, and all the difficulties and circumstances attending their navigation."[19] If there existed no chart of sufficient detail, Washington wanted a quick survey and sounding of the harbor at New London, a possible point of attack.[20]

The urgency of this request shows the paucity of Washington's cartographic resources. Not until July 1777 had the Americans provided for a small department of geographers under the command of a Scotsman, Surveyor General Robert Erskine. Erskine, a fellow of the Royal Society of London, was a businessman and inventor who had earned a scientific reputation for a new kind of pump. After four years in America he had embraced the patriot cause, but Erskine died suddenly in 1780 at age forty-five. Washington quickly promoted the Scotsman's assistant, Simeon De Witt. De Witt's inadequate force was never larger than twenty. Crippled for lack of training, the geographers disbanded in 1783, their mission absorbed elsewhere in the army until the War of 1812. Both the British and the Americans had attached their surveyors to headquarters under the commander in chief. Both, however, organized special units with some reluctance more than two years after the initial exchange of casualties on Lexington green.

Thus, the geographers on both sides of the conflict shared a frontier profession that mixed astronomy and natural science with strategic concerns. Some soldiers reached for science through scholarly classification: lists of insects and vegetation, fossil discoveries, studies of Indian tribes with linguistic information. Their maps reappeared in textbooks, guidebooks, and secondhand travel accounts. Dr. John Mitchell, a Virginia naturalist, borrowed from army surveys for his landmark *Map of the British and French Dominions in North America* (1755), the most complete map of its time. Meanwhile the British studies of American coastline set new standards for maritime surveying. William De Brahm, surveyor general of the Southern Department, explored Florida and shortened the passage to Britain with a chart of the Gulf Stream. Published in 1772 in De Brahm's great atlas, *The Atlantic Pilot,* it was followed by Captain Des Barres's aquatint engravings of the northeastern seaboard in *The Atlantic Neptune* (1777). Franklin also studied the Gulf Stream but his work remained a scientific curiosity that was generally ignored by sailors. Not until 1796 did Americans rival the British with *The American Coast Pilot,* the first in a series of important nautical atlases compiled in Massachusetts and New York by the publisher Edmund March Blunt.[21]

At first the British had focused on inlets and bays where surveyors could use heavy equipment that was hard to remove from the ship. By the 1770s the technology had followed the rivers inland through the work of wilderness scouts like Thomas Hutchins, one of Gage's resourceful surveyors. Hutchins belongs to the history of U.S. topographical engineering for two reasons. As the first "Geographer to the United States," 1781 to 1789, he promoted a useful method of long-distance triangulation that brought order and sophistication to federal surveys. He also anticipated the role of the soldier-surveyor as a versatile water expert. A river surveyor with a talent for self-promotion and a keen interest in the lush Mississippi heartland,

Hutchins was a transitional figure, a pioneer in the progress of frontier surveying from the austere reporting of hazard to the appraising of strategic resources in a wild and contested terrain.

An orphan from Monmouth County, New Jersey, Hutchins learned to survey in the Pennsylvania militia and the Royal American Regiment that conquered New France. Setting out for the western country before he was even sixteen, Hutchins crossed the Alleghenies at a critical time for the Ohio frontier. He reached the forks of the Ohio in 1758 when the French burned and abandoned Fort Duquesne. He was there the following year for the founding of Fort Pitt, future Pittsburgh; there in 1763 when the British took control of the Ohio Valley; there when Col. Henry Bouquet stopped Chief Pontiac at Bushy Run; there again when the new United States claimed the territory as its own in 1783. His river reconnaissance began as a scout for Bouquet and a famous deputy agent for Indian affairs, George Croghan. Bouquet and Croghan used the young surveyor to guide troops to Fort Niagara, map British conquests, spy, and negotiate for the release of prisoners. Scouting introduced Hutchins to rivers and science. It also gave him a taste of the intrigue that proved his undoing later in life.[22]

In 1766, after four dull years as an ensign and paymaster at Fort Pitt, Hutchins jumped at the chance to join Croghan and engineer Harry Gordon on an Ohio River expedition to the newly acquired territory of Illinois. Croghan had been west before. In 1756 he had explored to the Miami River with the Maryland scout Christopher Gist, but the Illinois expedition was Croghan's most elaborate venture, the largest flotilla yet assembled at the head of the Ohio. Hutchins, now a lieutenant, signed on as the surveyor. Embarking on June 18, 1766, the party proceeded without incident to the three miles of rapids, or "falls," near the future city of Louisville, Kentucky. Hutchins fixed latitude and sketched the rapids, pronouncing them "difficult." Elsewhere he recorded temperature, current velocity, mineral deposits, vegetation, crops, wildlife, even "the different nations and tribes of Indians, with the number of fighting men."[23] In August the party reached the Mississippi. Hutchins and Gordon explored north to the Illinois River, reversed course, then descended to New Orleans where they boarded a ship for New York by way of Pensacola and Havana. Hutchins then traveled overland to Fort Pitt, thus completing nearly the full circle of the awesome Anglo-American empire at the very peak of its influence and power.

In 1768 Hutchins again mapped the Ohio. While stationed in British Illinois he compiled surveys and exchanged notes with Philip Pittman. Pittman used Hutchins' sketch of the Kaskaskia River for his own publication. Hutchins, in turn, relied on Pittman's information for his map of the lower

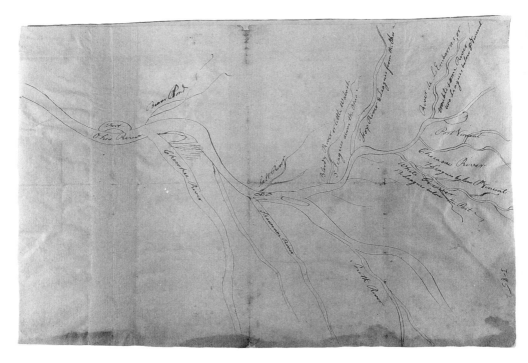

Thomas Hutchins's sketch map of the
Ohio at the mouths of the Tennessee and
Cumberland rivers (the "Cherokee" and
"Shawanoe"), about 1768. Courtesy of the
Historical Society of Pennsylvania.

Mississippi. After correcting the map, Hutchins added *A Topographical Description of Virginia, Pennsylvania, Maryland, and North Carolina*, first published in London in 1778. The book with its map and a "Plan of the Rapids" survived for almost a half century as a standard guide to the Ohio country. Meanwhile the surveyor edited his reports on Illinois for the American Philosophical Society. Toying with science, he froze mercury at Hudson Bay and tested surveying equipment. The Royal Society published some of his compass experiments in 1775.[24]

When the colonies broke with Britain, Hutchins was in London prosecuting a salary dispute with the Lords Commissioners of the Treasury. Suddenly, on August 29, 1779, the captain was placed under arrest. Hutchins later claimed that his only offense was the refusal to bear arms against his countrymen, but neither his actions nor his correspondence showed prior sympathy for the patriot cause. Sir John Fielding, a British magistrate, had another explanation: Hutchins had been passing secrets to American agents in Paris. The secrets were probably nothing more than financial reports on gyrating government stocks, but coded letters and a dummy mailing address made Hutchins look like a felon. His French connection turned out to be Samuel Wharton, a Philadelphia merchant. After a tense seven weeks in Clerkenwell prison, Hutchins, now broke and in frail health, was released for lack of evidence. Only then did he defect. "He is esteemed a good officer and an excellent engineer," said Franklin from Paris.[25] Escaping to America by way of France, Hutchins reached Philadelphia in early 1781. There, at the close of the war, Congress appointed him "Geographer of the Southern Army," then "Geographer to the United States."[26]

United States Geographer Thomas Hutchins began his patriotic career exactly where he had left off with the Lords Commissioners of the Treasury—haggling over assignment and pay. Not until April 1782 did he leave Philadelphia for the southern campaign. He served Gen. Nathanael Greene with little distinction but retained his lofty title and emerged prominent among the new American class of long-distance surveyors. Hutchins found great demand for his profession as a canal inspector, cartographer, and Congressional geographer at large. In 1784 he joined fellow astronomers Rittenhouse, Ellicott, John Ewing, and others on the Pennsylvania-Virginia line, possibly the most disputed of the colonial boundaries, certainly the most surveyed. With due ceremony the assembly proclaimed "an anxious desire to gratify the astronomical world in the performance of a problem which has never yet been attempted in any country, by a precision and accuracy that would do no dishonor to our characters."[27] The problem not yet attempted was a great triangulation, the fixing of latitude and longitude by sun and stars over a distance of several hundred miles. These same principles of astronomical triangulation allowed Hutchins to run an accurate "baseline" across the Seven Ranges of the Old Northwest. This early work on the Seven Ranges began the checkerboard pattern of federal townships, a system that survives to this day.[28]

Meanwhile the Hutchins maps and journals were gaining a wide audience in London, Dublin, Paris, Boston, Philadelphia, and throughout the frontier. Perhaps his most literary and scholarly achievement was *An Historical Narrative and Topographical Description of Louisiana, and West-Florida,* composed during the author's unfortunate stay in London and then published in Philadelphia in 1784. His focus was on navigation but Hutchins often digressed. In his report on the Mississippi, for example, he explained the pattern of flooding that seemed to be building the Delta. Hutchins believed that the land on both sides of the river had surfaced through the ages as silt flooded the lowlands when channels became jammed with debris.

Lakes on either side of the Mississippi appeared to be bays cut off from the Gulf. One day they too would be part of the Delta. "Thus," Hutchins concluded, "the land is annually raised and constantly gains of the sea."[29] A footnote cited Venice, the Seine, and other places where mud deposits were choking off navigation. Anticipating the nineteenth-century topographical engineers, Hutchins touched upon lasting concerns: specifically, the link between flooding and shallow deposits; generally, the attempt to understand American hydrology through European examples.[30]

The more immediate students of *An Historical Narrative* were civilian geographers at work on maps of North America. "It is with the greatest diffidence I beg leave to lay at your feet a very humble attempt to promote a science of which you are so bright an ornament," wrote John Fitch to Hutchins in the dedication to his *Map of the Northwest Parts of the United States* (1785).[31] Another admirer was Jedidiah Morse, the Connecticut schoolmaster who paid homage to Hutchins in the 1794 edition of his own great work, *The American Geography*. Morse, citing Hutchins, reported the West as network of magnificent rivers, a gateway, perhaps, to the unknown Stoney Mountains that bordered a Great South Sea. Meanwhile the Hutchins "Plan of the Rapids of the Ohio" reappeared in John Filson's book on Kentucky (1793). The Mississippi and Ohio journals were appended to the third edition of Gilbert Imlay's popular guidebook, *A Topographical Description of the Western Territory* (1797).[32]

Pirated and reprinted, the maps and journals of Thomas Hutchins contradicted the rumors of wasteland in upper Louisiana, reports of "an absolute barren [region] that nobody knows the bounds of or cares," a territory "sandy [as the] deserts of Africa."[33] Hutchins gave no indication of the western desert. He seemed to value Louisiana not for the promise of a northwest passage or gems or golden cities but for two more immediate reasons. First was the military control of the western rivers. "The free navigation of the Mississippi is a joke whilst the Spaniards are in possession of New Orleans," wrote Hutchins in an unpublished draft of *An Historical Narrative;* "the power and grandeur of that part of the United States on and westward of the Ohio . . . will certainly depend on our having possession of Orleans."[34] Second was Hutchins's conviction that "no country in North-America, or perhaps in the universe, exceeds the neighborhood of the Mississippi in fertility of soil and temperature of climate."[35] Control of New Orleans; open navigation to America's bountiful heartland—the connection grew strong in the mind of Thomas Jefferson, a close student of British surveys.

As mapmakers exchanged information and the lure of the great Mississippi inched Britain toward a showdown with Spain, Hutchins, sharing some

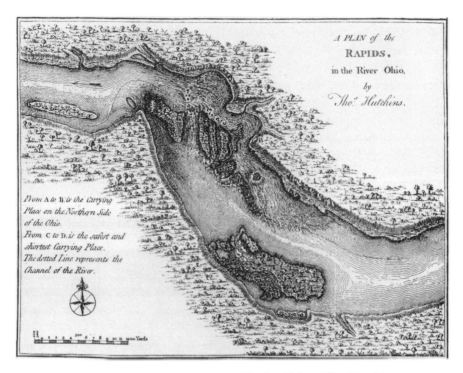

A PLAN of the RAPIDS, in the River Ohio, *by* Thos. Hutchins.

From A to B. is the Carrying Place on the Northern Side of the Ohio.
From C to D. is the safest and shortest Carrying Place.
The dotted Line represents the Channel of the River.

The "rapids" or falls of the Ohio at
Louisville, Kentucky, 1778. Courtesy of
the American Philosophical Society.

of Jefferson's vision, wrote about a "potent empire" of the interior, a frontier "copy" of England.[36] Pushing for a British raid on the Delta, he outlined
two possible lines of attack: a flotilla armed at Fort Pitt might take New Orleans from the north; or, if the enemy controlled the river, an army might
sneak up the back of the Spanish defenses from a beachhead at Lake Borgne.
It was a scenario for disaster. Almost exactly a half century later, as troops
slogged through the Delta against enemy fire, the British army suffered its
darkest hour at the Battle of New Orleans.

The Delta seemed worth the bloodshed because of the valley upstream.
Hutchins reported an abundance of corn, rice, silk, cotton, indigo, rhubarb,
hemp, and flax; also, the "richest fruits in great variety" and timber "as fine
as any in the world." In short, the explorer insisted, all that was "rich or rare
in the most desirable climate in Europe, seems natural to such a degree on
the Mississippi."[37] So impressed was Hutchins that he acquired thousands of
acres near Natchez and Baton Rouge, but he lost the riverfront property

when the Spanish took over the region. In 1788, about a year before his death, Hutchins tried to reclaim the land. According to letters between Hutchins and a Spanish envoy, the geographer, supporting a plot hatched by the devious mind of furtrader George Morgan, would map and help colonize Louisiana as the secret agent of Spain.[38]

Hutchins, then, was not a dedicated patriot, but he helped the United States understand the military and commercial importance of the complex Mississippi system. He was not a schooled theoretician, but with the precision of science he refocused frontier geography, moving from the search for El Dorado to a concern for waterway access to prime agricultural lands. His West became Jefferson's hope for a democracy of steady freeholders in vast Louisiana, a farming frontier. The region, no longer "barren," was now "fertile" and "benign."[39] It remained for future generations to sour that perception with reports of an American desert north of the Rio Grande.

2. "A Preference in Favor of London"

Hutchins died in Pittsburgh in 1789 just two days before the first president was sworn into office at New York's Federal Hall. Washington called him a man with a "taste for science in general," a pioneer of the frontier survey.[40] But Hutchins also had many detractors. His work on rectangular townships had been precise but painfully slow and hardly a ringing endorsement for government aid to science. The western delegation in Congress scoffed at triangulation. And despite Franklin's plea for federal canal investigations, that application of science seemed foreign to the British tradition—too despotic, too French. The U.S. Constitution mentioned science only once: Congress, like the British Crown, would have the power to "promote Science and useful knowledge" through the granting of patents.[41] Soon the Federalists were looking to London for a system of weights and measures that extended the nation's dependence on British surveying equipment. Even the Federalist army, Madison feared, was adopting "the Engl[is]h standard."[42]

Topographical engineering reinforced the British·tradition at a time when the fort agency at West Point still yearned for contact with France. The first U.S. surveyors were American, Swiss, and French as well as Scottish and English, yet four men who launched the mapmaking program—Erskine, Hutchins, William Tatham, and Isaac Roberdeau—were British in many respects, all having worked in London. Robert Erskine, educated at the University of Edinburgh, had come to America from London as an agent

William Tatham—surveyor and author, the founder of an army topographical office during the War of 1812. Courtesy of the North Carolina Division of Archives and History.

for British investors. Faith in the Yankee mechanic fed Erskine's belief that "young gentlemen of Mathematical genius" could become "expert survey-ors" with "a few days practice."[43] Hutchins, American by birth, had re-treated to London in 1776, and there as a British captain he had revised the *Topographical Description* and the *Historical Narrative,* his two great works. The Englishman William Tatham had also studied in London as did the first chief of the U.S. Topographical Bureau, Isaac Roberdeau.

It was the least educated of the four who showed the greatest "prefer-ence," as he put it, "in favor of London as the focal point of emulation for operations of magnitude in the whole world of [internal] *improvement, com-merce,* and *finance,* whether in the kingdom of England, or the remotest cor-ners of the earth."[44] The hyperbole typified its author, William Tatham, the prolific dreamer of countless schemes for inland navigation—an intracoastal waterway, a federal department of public works, gunboats, steamboats, a marine railroad over the falls of Niagara. Tatham had immigrated in 1769 as apprentice to the mercantile house of Carter and Trent. Fortune eluded him at every turn, as a tobacco trader, soldier, surveyor, lawyer, state legislator, and self-appointed diplomat on a mysterious mission to Spain. In 1790 he or-ganized a geographical department for Virginia, compiling *A Topographical Analysis of the Commonwealth,* but this too went sour when the governor ac-cused Tatham of misusing the state seal. Returning to London in 1796, he embarked on yet another career, a literary one, composing numerous unsuc-cessful works on agriculture and civil engineering.[45]

Water-powered inclined plane for raising
boats, from William Tatham's *Political
Economy of Inland Navigation, Irrigation,
and Drainage* (London:n.p., 1799).
Courtesy of the Library of Congress.

Tatham, however, could recognize genius in others. In London he began
to collaborate with one of America's great inventors, Robert Fulton. Tatham
shadowed Fulton. When Fulton set down his ideas on world peace, Tatham
followed suit. When Fulton turned to the more limited objective of hoisting
boats over mountains, Tatham countered with a weighty treatise, *The Political Economy of Inland Navigation, Irrigation, and Drainage: With Thoughts
on . . . Construction of Canals.*

Published in 1799, *The Political Economy* was an artless paraphrase of Fulton, John Phillips, and other promoters of British canals. From a simplifica-
tion of classical economics came the notion that the wealth of nations was
somehow proportionate to their ability to transport goods. From James
Brindley, whom Tatham considered "the greatest natural mechanic, perhaps,
that was ever born," came the preference for artificial waterways over works
in natural channels.[46] It was Fulton, however, who suffered most in tran-
scription. Where the American had proposed a railroad for pulling boats
through the elevations of a canal, Tatham suggested an escalator powered by
the force of falling water. "Had it been intended to run Mr. Fulton's ingenious plans into a species of burlesque, it could scarcely have been carried

higher than this flighty scheme," said London's *Monthly Review*. As for Tatham's other proposals, they possessed "neither originality nor judgment; and . . . by applying the inventions of others (good within certain limitation) to a ridiculous excess, do the greatest possible injury to the spirit of forming canals."[47]

In America, however, imitation served a purpose. Returning to northern Virginia in 1805, Tatham continued to demonstrate a spongelike ability to retain information and this won him friends in Washington City. "We never expect from the writer a detailed, well-digested and practical plan," Jefferson reflected in 1807, "but good ideas and [those] susceptible of improvement sometimes escape from him."[48] One of his more prophetic presumptions was the need for a geographer's department, himself in command. The department would provide "a national depot" of maps and manuscripts, "of geographic, topographic, ichnographic, and hydrographic surveys; of military designs, plans, sections, elevations, views and working drafts."[49] Tatham had brought with him from England just such an archive for Congress. It included some valuable items—a set of military journals, plans for taking Quebec and the defense of Washington, maps of Florida, the West, and the Canadian frontier. One writer says Tatham's proposal for a cartographic archive was a seed of the Library of Congress.[50]

Tatham had other ideas. Using Indian labor, he would lay pontoons across the Great Lakes. He would install a telegraph, fortify the Potomac, and copy the London dry docks for the U.S. Navy. At last in 1812, as American militias unraveled at Detroit and Montreal, the War Department took interest in Tatham's Canadian maps. Madison thought it might be wise to purchase the entire collection, but the House tabled the plan. Tatham persisted. Finding an apartment on F Street close to the Treasury Building, he began to copy valuable maps. It was the strength of his collection rather than the collector himself that prompted the War Department to retain Tatham as a draftsman for about $100 a month. In July 1814, the department upgraded his status. He was now a civilian "topographical engineer," a caretaker of instruments, journals, and plans. Soon he was calling his tiny office the "topographical department," even as the army was organizing a separate "topographical section" of combat surveyors. Tatham made the most of his appointment by spying on the British army and racing through Washington City with news of the impending attack. Then as Washington burned he escaped up the Potomac with government papers and maps. At age sixty-two the vagabond engineer was finally proving his worth.[51]

Flexible, expendable men like Tatham were a link to British surveying that laid the archival foundation of the future topographical bureau. Their stories make a point about the transatlantic migration of topographical engineering: innovation, independent of science, spread widely through the efforts of minor officials who traced plans and surveys. Tatham, an emulator, was chiefly employed as a man who copied information. A map and book collector, he wrote extensively on hydraulics, and he stressed the importance of rivers in a flood of borrowed ideas. In 1813, for example, he petitioned Congress with a long passage from the writings of Gen. Henry ("Light-Horse Harry") Lee. "Government," said Lee, "ought to provide, in time of peace, maps, on a large scale, of the various districts of the country, designating *particularly* the rivers, their tributary streams, the bridges, morasses, and defiles." Tatham then blamed the fall of Detroit on the army's inferior knowledge of frontier river systems. Shaming Congress, he said the United States was perhaps the only "enlightened" government without a corps of army surveyors.[52]

By the time Tatham's letter reached Congress, the army had already appointed about eight topographical officers, but they never worked as a single command. One of the first appointments was Maj. John Anderson of Vermont. An early West Pointer, he had been under siege with Gen. William Hull when the Americans surrendered Detroit. A second topographer from West Point was a major with legal training, J. J. Abert. The one French recruit was a major who had served with Tousard in the old corps of artillerists and engineers. All were honorably discharged in 1815. Tatham was fired. Not until the era of John C. Calhoun did the War Department provide for a topographical bureau as it consolidated the army under a general staff.[53]

As surveyors searched for a place in the army, Chief Engineer Joseph Swift sent three men shopping in Europe. Two purchased books in Paris. The third, James Renwick, was an English-born scientist and engineer who, like Tatham and Erskine, specialized in hydraulics. Renwick traveled through France, England, and Scotland with a list of instruments for West Point—a centrifugal pump, a water distillery, an apparatus for weighing objects in water, a bottle for weighing air, a copper air fountain, hydrostatic bellows, and a set of glass balls for gravity experiments. He also wanted some expensive surveying equipment like a silver azimuth circle. Although he allowed that the French savants were "the boast of the age," Renwick ordered the equipment in London, noting "the superiority of England over the nations of the continent" in things apart from science.[54] British instruments found their way to the new topographical office in Georgetown, the army's archive and storage facility. When Karl Bernard, a visiting duke, inspected the War Department in 1825, he was surprised to find British-made telescopes, theodolites, repeating circles, two transit instruments for a proposed observatory,

Isaac Roberdeau, first chief of the Corps'
topographical bureau, 1817–1829. Courtesy
of the Historical Society of Pennsylvania.

and an English yardstick. Bernard noted that Americans were buying from
London at a time when Holland and other nations were adopting the standard of weights and measures pioneered by the French, the superior metric
system.[55]

The man who cared for the army's mostly British collection was the major
now referred to as "chief" of the topographical office, Isaac Roberdeau. Few
Americans could match his credentials. Schooled in Philadelphia, he had
studied engineering in London and inspected the great canals. In 1791 and
1792 he had worked for Pierre L'Enfant and Andrew Ellicott on their plan
for the federal city. He later surveyed for L'Enfant at Paterson, New Jersey, a
planned industrial center on the Passaic River, America's first "company
town."[56] Roberdeau's canal work began on the Schuylkill link to the Susquehanna River, one of William Weston's projects. As a major during the
war he had surveyed fortifications and taught topography at West Point.
Soon after the Treaty of Ghent, Roberdeau joined topographer Anderson on
the St. Lawrence River to complete the boundary line between the United
States and Canada, a nine-hundred-mile survey.[57]

French ancestry and family connections may have helped the surveyor's
career. The eldest son of congressman and wealthy French-American Daniel
Roberdeau, Isaac was born into educated society, and he rose in government
service on the goodwill of powerful men such as John C. Calhoun, John Q.
Adams, Madison, and Lafayette. His sister married the uncle of Williams's
successor, Brigadier General Swift. Swift had been a young hellion at the
military academy before rising through the ranks to chief engineer. In 1809,

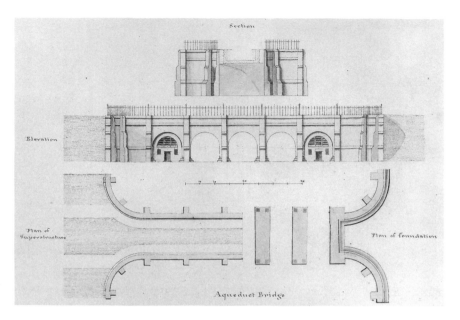

Plan of an aqueduct, from Isaac
Roberdeau's "Mathematics and Treatise
on Canals," around 1796. Courtesy of
the Library of Congress.

when Roberdeau was looking for work in the fortification program, Swift
had introduced the surveyor to Secretary of War William Eustis. In 1815 the
chief engineer brought Roberdeau to West Point, and Swift later saved him
a choice position as an assistant to Major Totten and General Bernard on the
U.S. Board of Engineers for Fortifications. In August 1818, the major was
called to Washington as part of the general staff. Roberdeau, now fifty-four,
received his next assignment: "I was then ordered to take charge of the formation of a Topographical Bureau, upon the plan adopted and pursued in
France by [Lazare] Carnot."[58]

As much as the high command may have valued French engineering,
Roberdeau was a field-trained surveyor, his methods more British than
French. A clue to this orientation survived in Roberdeau's useful "treatise"
on mathematics and the Pennsylvania canals. Composed between 1792 and
1796, the book endorsed James Brindley's preference for aqueducts that kept
canals from merging with rivers. Elsewhere it praised the Englishman Weston
for disregarding French attempts to standardize canals by computing the

slope or "angle of inclination" of the banks of a canal. Slope, Roberdeau agreed, should be "deduced from experiment" and determined "independently of all mathematical reasoning on the properties of angles." Standard formulas seemed "wholly irrelevant" to canaling in Pennsylvania.[59] Years later Roberdeau's British training resurfaced in an essay that tried to define topographical engineering: topographers, the major explained to Congress, were like English civil engineers, only native born and more specialized. A third example of British thinking was Roberdeau's challenge to the Treasury Department for control of the coast survey. In Britain, the major insisted, the best surveyors were army and navy officers who studied the coastline section by section in topographical brigades. Not until 1834, five years after Roberdeau's death, did topographers try to link the bureau to French tradition with vague comparisons to the Corps des ponts et chaussées.[60]

Cartographer and canal expert, the grandson of a Frenchman yet a link to British surveying, Roberdeau brought science and sophistication to his loose and varied command. Three of the ten topographical engineers were early West Pointers. Three had been trained in Europe, two in France. Maj. James Kearney, an Irish immigrant, had once been a draftsman for Tatham. There was also a lawyer, a fort specialist, a schoolmaster, and a future diplomat— together a talented lot, yet widely dispersed. Capt. Hugh Young joined Andrew Jackson's army for the march into Florida. Kearney surveyed Dismal Swamp. Anderson served with ordnance; Abert, with the coast survey. This left Roberdeau alone in Georgetown with nothing to command and little to do. He compiled a map of the West, sponsored astronomy projects, and planned a canal past the White House. He discussed science with John Quincy Adams. He supported the presidential ambition of his good friend Calhoun. In February 1821, the House Committee on Military Affairs censured Roberdeau for receiving an extra $1.25 per day for no apparent reason. Secretary Calhoun rose to the major's defense, but the stigma remained. "He has no superintendence whatever of the operations of the officers of his own corps, no means of obtaining a knowledge of their employ, no influence in their destination or duties," Abert reported. "To call one so situated, in charge of the Bureau, is a mistake in terms."[61]

The bureau was further distracted by shifts in the Corps' command. Chief Engineer Swift had pushed for topographical engineering, but when he resigned in November 1818, the bureau was just taking shape. Swift's successor, Col. Walker K. Armistead, was a fort builder and artillerist who ignored and weakened the bureau. Then in 1821 the army promoted Col. Alexander Macomb, the third chief engineer in three hectic years. "In all regular establishments," Macomb believed, "it is necessary and convenient that there should be order and form, and that there should be different grades in the officers, suitable to the circumstances of the service to be performed."[62] Macomb admitted the bureau had neither—no order or form, no sense of the vital service the surveyors would have it perform.

3. The Long Expeditions

The Corps' topographical bureau searched for that sense of purpose in the fame of its first explorer, Maj. Stephen Harriman Long. Historians know Long as a railroad surveyor, bridge engineer, steamboat inventor, and the "foster parent" of government exploration, the man who revived the myth of the desert on the Kansas-Nebraska plain.[63] A Dartmouth graduate, he had worked as a math teacher and a school principal before Swift advised him to go to New York to plan fortifications. Swift called him "a gentleman of large mechanical ingenuity," an apt assessment, for Long was more mechanic than scholar although professionally something of both.[64] Long, like Thomas Hutchins, was a savant of a practical science that thrived in a middle ground between overlapping professions, probing the strategic connection between frontier rivers and frontier defense.

To the gallery of portraits might be added another: Long was a water expert—a pilot, a dam builder, the first topographer to do more than study the

Stephen H. Long, by Charles Willson Peale, 1819. Courtesy of the Independence National Historical Park Collection

hazards to steamboat navigation. This interest surfaced in August 1816 as Long followed the Illinois River to desolate Fort Clark. The fort was "illy calculated for defense," but the rivers held promise.[65] The Saginaw, Wabash, Kankakee, and St. Joseph might all be improved at small expense, wrote Long to the War Department. A canal between the marshes of the Illinois and the Chicago River, a mere two miles, would allow freight to pass from the Mississippi to the lakes. The Chicago could become "a safe and commodious harbor" by sinking jetties or piers on either side of its entrance and dredging the bar between. Elsewhere, said Long, "nothing more is necessary than the construction of sluices in a few places where there are ripples [small rapids]."[66] Another canal might connect Lakes Michigan and Erie by way of the Maumee; still another, the Mississippi and Ohio just above their junction. It was a sensible plan. The Illinois-Michigan canal bill, passed by Congress in 1827, granted land for excavation along the same route proposed by Long. Topographers resurveyed the site two years later, although more than a decade passed before construction began in earnest. In 1848 the waterway opened at last.[67]

Meanwhile back in Missouri, at the garrison near St. Louis called Belle Fontaine, Long proposed a grand reconnaissance in a boat of his own design. "I would build a small steam boat about 40 feet in length and 7 feet beam, drawing no more than 14 inches of water," wrote Long in a letter to President James Monroe, March 15, 1817. "I would navigate all rivers of consequence falling into the Mississippi, meander their courses, and take the latitude and longitude of their mouths, and heades of navigations."[68] At the time, there were those in the army who argued for more substantial gunboats—warships to impress the natives, boats large enough for a company of troops. Long preferred a smaller and shallower boat with a crew of West Point cadets. Monroe, however, did not respond, perhaps unwilling in 1816 to risk his frontier policy on an unproven machine. Yet the shallow-draft army steamboat, like Long's western canals, was a technology on the near horizon just a few years ahead of its time.

Instead of a steamboat the major was given a six-oared skiff, a nine-man crew, and a northern assignment. He would navigate the Mississippi to the mouth of the Wisconsin and follow that river east through Winnebago country. Crossing to Lake Michigan, he would inspect fortifications as far north as Green Bay. On June 1, 1817, Long set off from Belle Fontaine. After three weeks on the river, the party reached Prairie du Chien, Wisconsin, and there it rode overland to the Fox-Wisconsin portage where Long reversed his course. Exactly why he stopped short of Green Bay has not been recorded. Perhaps he had sufficient intelligence from settlers, or perhaps his interest was drawn farther north by the Falls of St. Anthony, a spectacle recorded by Pike. Returning to the Mississippi, Long reached the falls at the head of the river on July 16. The site was "a beautiful cascade . . . the most interesting and magnificent of any I have ever witnessed," but the euphoria ended abruptly when Long assessed his supplies. Food was low. The whiskey, "a necessary of life for those who navigate the Mississippi in hot weather," was already dry.[69] Fighting rain and dangerous currents, Long made haste for the Missouri. He verified measurements, took a second look at three Mississippi forts, recorded some Indian lore, and narrowly escaped the rapids at Des Moines before the explorers returned to Belle Fontaine.[70]

Long had covered about eighteen hundred miles in seventy-six days. The journey's scientific value, however, was difficult to assess. There were no shattering discoveries or new understandings with the Indians. Long had mismanaged provisions and ignored, seemingly, his primary objective of crossing to Lake Michigan by way of Green Bay. In the tradition of the great explorers the major had worked on a journal that strived for poetic description but confused some of the facts. "Mississippi," the Algonquion Indian word for "Great River," became "Clear River" in Long's account. He guessed at the condition of the Black River and overestimated the length of its navigation by at least sixty miles. For the most part, however, Long confined his narrative to what could be seen from the boat—the sandbars, rapids, climate, vegetation, and strata of rock.[71]

Long's 1817 journal reflected the wandering focus of army investigations too narrow for the general reader yet too common for natural science. Intended for a popular audience, it remained unpublished until 1860 when the Minnesota Historical Society rediscovered this neglected account of the skiff expedition. Meanwhile Long put down some thoughts on the geological origins of the upper Mississippi. Three scientists at the American Philosophical Society acknowledged that Long, "a reflecting and intelligent mind," had brought back "valuable information," but they gently refused publication.[72]

Long had more success with terse reports that assessed navigation. In a letter to Brig. Gen. Thomas A. Smith, the commander at Belle Fontaine, Long evaluated the rivers of empire that took trappers into the wild. The Missouri, said Long, was a powerful river with shifting currents that were difficult to sail. South of the Platte at the base of the Rockies were "numerous Prairies" and "extensive sandy deserts" cut off from navigation, but six hundred miles of the Arkansas seemed deep enough for keelboats, even in summer.[73] The St. Croix was broken by rapids. Summer sandbars blocked the Wisconsin River, and the Fox to Green Bay had dangerous shoals. Turning south, Long reported the "great raft" of the Red, a jungle-like mass of logs about five hundred miles northwest of Baton Rogue. At flood stage a keelboat could pass the raft and travel far up the river at about sixteen miles a day.

Long, like Hutchins, was careful to rate each river's importance to frontier defense. The explorer thought forts at Chicago and Green Bay should be formidable enough to secure the two most promising lines of river communication between the Mississippi and Great Lakes. A post near the Falls of St. Anthony, the future site of Minneapolis, would command the Mississippi headwaters and intercept British trappers. Farther west he suggested a string of forts along the Missouri: one on the Platte near present-day Omaha, another 570 miles beyond in the present state of South Dakota, a third in the Mandan nation "in order to controul the trade above and prevent an enemy from cutting off necessary supplies," and a fourth at the mouth of the Yellowstone in future North Dakota.[74] Lewis and Clark had inspected most of these sites. Fort builders at the Falls of St. Anthony had already purchased the land. If Long made a real contribution, it was one of synthesis and corroboration. Endorsing the need for a Yellowstone expedition and supporting men like Calhoun, Long viewed the West as a system of river defenses that would block the British claim to the contested Northwest.

On December 11, 1818, the War Department proposed a line of forts from Chicago to Green Bay to the Falls of St. Anthony and west along the Missouri from the Mandan villages to the mouth of the Yellowstone—the same sites endorsed by Long. Soon the army was planning the 1819 Yellowstone expedition, a famous fiasco. Colonel Henry Atkinson would penetrate the upper Missouri with up to eleven hundred men. Major Long and a party of Philadelphia scientists would precede the expedition, mapping its route. Reviving his steamboat idea, Long would build a shallow-draft stern-wheeler, the first of its kind. He would clarify international boundaries. He would sample, survey, and otherwise record "every thing interesting in relation to soil, face of the country, water courses and productions whether animal, vegetable or mineral." [75]

Spectators cheered and a cannon fired when the scientific wing of the Yellowstone expedition arrived in Pittsburgh on May 3, 1819. With Long in command, they numbered twenty-four—a pilot, clerk, carpenter, and steam engineer; also thirteen soldiers, four scholars, two cabin boys, and a landscape artist from Britain. Their boat was an experimental stern-wheeler of Long's own design. A cam cutout allowed the engine to operate on full or partial steam, another innovation. Mounted with cannon, its boiler hidden from view, the machine was "a huge serpent, black and scaly, rising out of the water from under the boat, his head as high as the deck, darted forward,

his mouth open, vomiting smoke, and apparently carrying the boat on his back."[76] Long called it the *Western Engineer.* When the steamboat reached Fort Lisa on September 17, it was the first to navigate the Missouri beyond Chariton, nearly twice as far as the lead steamer of Atkinson's flotilla still mired downstream.

Despite this achievement, and despite the major's initial report to the contrary, the *Western Engineer* was a disappointment to Long and Calhoun. It had averaged six to eight tortuous miles a day from St. Louis to Fort Lisa, no faster than the common keelboat. The yellow Missouri had been a navigator's nightmare of bars and snags and silt thick enough to choke out the boilers on a regular basis. Dr. James Baldwin, botanist, detested the voyage. Crippled by tuberculosis, suffering loudly, it was all the doctor could do to stagger ashore at Franklin, Missouri, where he resigned, then died the following month. The others continued to a sandy cove above the Platte, their ordeal with steam at an end. Long traveled east without the *Western Engineer* while the military wing of the Yellowstone expedition, now about three hundred troops, continued to struggle upstream.[77]

That winter nearly a third of Atkinson's force would die of scurvy, exposure, and starvation at frigid Council Bluffs. It was in this season of panic and scandal that Long received a new, more reckless assignment—to survey the Platte, Arkansas, and Red, all at their mountainous source.

Long ventured overland with a reorganized escort in the spring of 1820, heading west from the Missouri to the mountains, south across wasteland real and imagined, through trials of theft, desertion, near-mutiny,

near-starvation, down the unknown Canadian River to the Arkansas and familiar Fort Smith. The expedition had crossed from what is now Nebraska to the Colorado Rockies then back through Kansas, the Texas Panhandle, and Oklahoma. "In regards to this extensive section of country," wrote Long of America's future breadbasket, "we do not hesitate in giving the opinion, that it is almost wholly unfit for cultivation, and of course uninhabitable by a people depending upon agriculture for their subsistence."[78] There were smaller findings as well. Gathered at Philadelphia in the autumn of 1821, the explorers displayed samples and at least 270 drawings of indigenous alpine flora, new species of wolf and coyote, fossils, insects, and what the *North American Review* generally considered "highly important additions" to geography and natural history.[79] Although Congress failed to publish these findings, the botanist Edwin James compiled his own now famous *Account of an Expedition from Pittsburgh to the Rocky Mountains,* printed in Philadelphia and London in 1823. Here Long achieved a fame that did more for his reputation in Europe than it did to impress the Corps.

Today anyone with a map of the Great Plains can see what the explorers attempted and where they went wrong. James, climbing with four companions, had scaled Pikes Peak for the first recorded view of the basin beyond. To his left he sighted the Arkansas, to his right the Platte, and ahead to the west what was thought to be the Lewis fork of the Columbia. It was actually a branch of the Colorado. Long continued south while a splinter expedition wandered past the mouth of the Cimarron and failed to record it entirely. The main party mistook the Canadian River for the Red River. Then Long

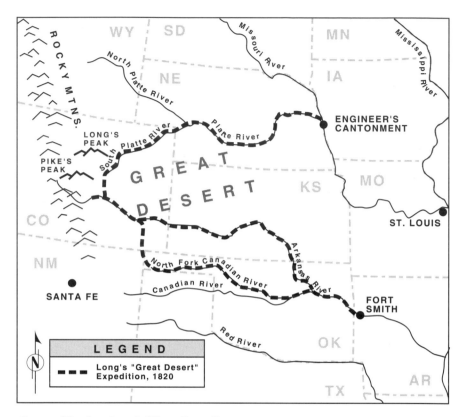

Route of Stephen Long's "Great Desert"
expedition, 1820. Based on S. H. Long,
*A Sketch of the Country Comprehending
the Line of Demarcation Between the Spanish Provinces and the United States Territory, 1820,* Record Group 77, Cartographic
Division, National Archives.

headed east before the explorers had located the headwaters of any important tributary. Even so, the expedition's "Map of the Country from the
Missouri River to the Rocky Mountains" was enthusiastically received. Compiled in 1822 by Long and Lt. William H. Swift, General Swift's younger
brother, it was the first delineation of the Arkansas-Canadian system. But the
map contained misinformation as well, enough to revive the ancient notion
of a single, mountainous portage between rivers of the Pacific and the Gulf.[80]

The larger problem has been attributed to oversight rather than error.
James proclaimed the prairie a "dreary expanse of almost naked sand," while
his companion, Capt. John Bell, echoing the explorer Pike, compared the

"dusty plain of sand and gravel" to the vast Arabian desert.[81] Printed across Long's initial map was "GREAT DESERT," which became "GREAT AMERICAN DESERT" in a Philadelphia engraving and the popular Tanner atlas of 1823.[82] The desert image lent drama to the literary stereotype of the west as the hostile unknown—treeless, pathless, the limit of migration. "The meager herbage of the prairie promised nothing in favor of a hard and unyielding soil," read James Fenimore Cooper's novel of the west, *Prairie* (1827). Where Long discovered "a broad plain, unvaried by any object on which the eye can rest," Cooper envisioned "those broad plains, which extend, with so little diversity, of character, to the bases of the Rocky Mountains."[83] The desert endured in Cooper's later work *Nations of the Americans* (1828), also in the writings of Francis Parkman and Washington Irving, an adventure of Sherlock Holmes, and the desolate setting of L. Frank Baum's satirical classic, *The Wonderful Wizard of Oz* (1900). "When Dorothy stood in the doorway and looked around she saw nothing but the great grey prairie," wrote Baum, a native of New York State. "Not a tree nor a house broke the broad sweep of flat country . . . burned . . . sun blistered . . . dull and grey as everything else."[84]

In Oz and elsewhere the desert became America's symbol for barren civilization. Subsequent army reports, extending the desert concept to arid Texas, predicted an end to the cotton kingdom on the southwestern plains. In 1828 Congress set aside portions of that region for a great reservation, a land of exile for the Five Civilized Tribes. Long's impact on white migration remains open to speculation. Few westerners would have read the novels of Cooper, and fewer had access to army reports, although the word "desert" on maps may have discouraged settlement. Long knew the findings of science had political implications. A native of wet New Hampshire, Long preferred a forested landscape, and he assumed that land far from rivers would be impossible to farm. But the prairie had strategic importance. Long called it "a barrier" that might contain a sparse population and protect the nation from sudden attack.[85]

A clearer map of this line of defense was a goal of Long's last expedition, a second trip to the Sioux and Chippewa country of the disputed Northwest. Organized in 1823, the mission was vaguely scientific. Long and four scientists would follow the St. Peter's River (the Minnesota) to its source somewhere beyond Lake Superior. Leaving Philadelphia in April, Long reached the Canadian line with thirty-one men—naturalists, interpreters, a detachment of troops from Fort St. Anthony, the son of fort commander Josiah Snelling, the cousin of John Calhoun, and a rival explorer from Italy, Giacomo Beltrami. After surveying Pembina, a small trading outpost on the

Bison along the
Platte River, 1820.
Courtesy of the
Library of Congress.

forty-ninth parallel, Long hoisted a flag, fired a salute, and claimed the
settlement for the Union. He then followed the Red of the North into
Lake Winnipeg, Manitoba. There, reversing course, the party floated the
Winnipeg River to Lake of the Woods, descending through a glacial region,
"one of the most dreary imaginable," and along the northern shore of Lake
Superior to Sault Ste. Marie.[86] The trip, some forty-five hundred miles,
ended in New York City on October 25. Long and geologist William H.
Keating continued to Philadelphia, where they compiled a map and edited
the journals for the *Narrative of an Expedition to the Source of the St. Peter's
River, Lake Winnepeek, Lake of the Woods, &c.,* privately published for Keating in 1824.[87]

Keating's *Narrative* identified two classes of wasteland on this northern
frontier. One stretched from Mexico to Canada and bore "a manifest resemblance to the desert of Siberia" that none of the explorers had actually
seen.[88] The other was a watery waste—a torn, bug-infested landscape, "exceedingly rugged and broken, . . . its products so limited and meager, that it
seems never to have been claimed as a residence either by man or beast."[89]
Here and there lay fertile patches, but since the rivers were all but impass-
able, the land was again dismissed as a barrier to settlement beyond America's sphere.

Just as Hutchins had mixed geology into reports on the western rivers,
Long borrowed the language of science to make a strategic point. "The
hydrography of this [north Minnesota] region is as yet very defective,"
Long explained to Calhoun. Although the basin of the Rainy River could
be crossed "in a thousand directions," the descent into Lake of the Woods
was obstructed by "the shape, magnitude, and position of innumerable
lakes," "myriads of islands," and "the courses, sinuosities, and declivities of

countless channels by which they are united." The territory was "a wilder-ness of lakes, islands, and peninsulas; a mazy waste, so inhospitable and ir-reclaimable, as to mock the art and enterprise of man, and bid defiance to his industry."[90] Again the distribution of rivers and lakes—a region's hydrog-raphy—was the army's critical measure of the land's strategic worth.

Questions remain. Why did Long depreciate the potential of the northern Red River Valley, a rich lumber and mining frontier? How could the Spanish believe in the agricultural promise of arid California at a time when the plains at the base of the Rockies, equally fertile, were written off as barren land? Some say Long simply made errors. Historian G. Malcolm Lewis at-tributed the myth of the desert to "ignorance" and the "exaggerated ideas" of men poorly trained to record what they saw.[91] Others say Long was true to a conceptual science that worked backward from theory to facts. Histo-rian William H. Goetzmann, writing on science and the frontier in 1966, said the explorers were "programmed" to interpret the West through the narrow lens of their own expectations.[92] Hutchins saw the West as a garden. Lewis and Clark thought the grasslands of the western prairie were "fertile in the extreme."[93] But Long expected to find a desert. He assumed, rightly per-haps, that an army needed water to move men and supplies, that the strate-gic points of the West were the forks of major rivers, that a region cut off from the Mississippi-Missouri system would be difficult to settle and near impossible to defend.

Thus the rise of hydrographical science expanded the surveyor's knowl-edge as it narrowed his scope. Long, more cautious than Hutchins, was drawn to define "the boundary which nature seems to have fixed as the west-ern limit of our population," the head of navigation. Wooded or barren, arid or not, Long and his colleagues devalued isolated pockets of land. Even the copper deposits near Lake Superior were quickly dismissed as a "mere addi-tion" to science in Keating's final report.[94] Far from the safe Mississippi, the copper wealth of the north seemed fenced off by wild terrain. Keating pre-dicted it would take at least one hundred years for Americans to get at the metal.

In the summer of 1824, as Keating compiled the Red River reports, Long was back in the West on a another river assignment, moving from water sur-veying to water construction at Henderson Island, Kentucky—the first fed-eral dam. That year Henry Clay campaigned for navigation improvements, the U.S. Supreme Court confirmed the power of Congress to regulate river steamboats, three topographical brigades studied canal proposals, and Rob-erdeau, still in Georgetown, said a native force of transportation planners could wean the army away from France. Hydrographical exploration now

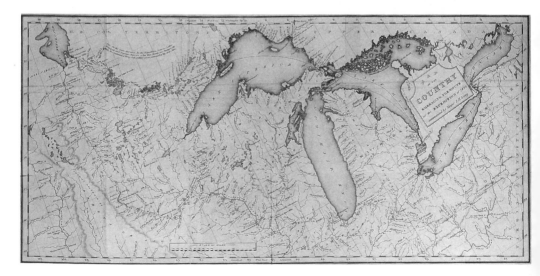

Route of Stephen Long's Northern Lake expedition. Based on maps in Roger L. Nichols and Patrick L. Halley, *Stephen Long and American Frontier Exploration* (Newark: University of Delaware Press, 1980), and William H. Keating, *Narrative of an Expedition to the South of St. Peter's River, Lake Winnepeek, Lake of the Woods, Etc. . . .* (Philadelphia: n.p., 1824).

Stephen Long's *Map of the Country Embracing the Route of the Expedition of 1823*, showing the inhospitable "mazy waste" of the rapids and falls west of Lake Superior. Courtesy of the American Philosophical Society.

served a double purpose. Professionally, it bonded the army surveyors to naturalists and others who, like Jefferson, looked for America's greatness on the exotic frontier. Water surveying was also the mission that focused the topographical bureau. Through the line of army tradition that connected Hutchins and Erskine to Roberdeau, Tatham, and Long, the bureau was laying claim to a vast jurisdiction. Surveyors were now the builders who used the tools of science to appraise the value of rivers and harbors on remote public lands.

4. Surveying—an Interpretive Tool

In the checkered achievement of Stephen Long was something of the topographer's future and something of his past: the seed of future support for military exploration; the stigma of a frontier profession still striving for scholarly status. Long had shown a return on a small investment in science. He had explored about twenty-six thousand miles, more than Lewis and Clark. He had discovered a mountain (Long's Peak) and a river (the Canadian) yet received no special commendation. Expecting a promotion, he was slapped with a bill for expenses. And although he had struggled for recognition at the American Philosophical Society, earning membership at last in 1823, he remained a topographer outside the inner circle of fort engineers. "We have the highest respect for all the corps of our little Army and we have much pride in believing that a more excellent body of officers is nowhere to be found," wrote Totten, ever patronizing the half engineers; "we have much pleasure in making most of them amongst our personal friends, but we see no where among them any, *not a single individual,* . . . whom we could see introduced into the Corps of Engineers without mortification of pain." [95] By 1825 the sentiment was mutual. "I believe they cordially hate us," Long told Roberdeau. Either that or builders like Totten were "jealous of our rising reputation." [96]

Or perhaps there was a deeper source of the tension between classes of army experts—a technological rift. Hutchins, Erskine, Tatham, Roberdeau, and Long all dabbled in conceptual science, yet all were also empiricists in the British tradition of the instinctive mechanic, men who learned chiefly by observation, driven experimenters who tinkered with machines and compensated for the lack of engineering degrees by testing scientific equipment. Long, the son of a cooper, began his engineering career as a teacher who toyed with hydraulics. His maps showed a delicate talent for instrumentation, and in 1820 the Long expedition, borrowing a technique from Europe, found a new way to measure the height of the Rockies by barometric pressure. Each time he returned from the West, Long, as he put it, resumed "the cause of Science" through work on clever devices.[97] In 1817 and 1819 he promoted his shallow-draft steamboat. After the Red expedition, Long invented a seesaw pulley system for lifting canal boats, a mechanical lock.[98]

Flag-waving dislike for the exiled French reinforced the British tradition. Frustrated with the Corps, Long competed with the Frenchman Crozet and others for an engineering job in Virginia. In February 1823, the appointment went to Col. William McRee, a gifted West Pointer. Long, although disappointed, was consoled not to lose to Crozet: "It affords me no small gratification to learn that I have been supplanted by a worthy American rather than a foreigner."[99] A year later Long and Roberdeau lobbied to remake the board of internal improvement commanded by the Frenchman Simon Bernard. Bernard's translator and aide was topographer Guillaume Poussin, one of the last of the French emigrés. Poussin later wrote an important book about the army and engineering. Stressing the "English origin" of the democratic republic, the French topographer said Americans respected the continental Europeans, but mechanical ingenuity and the hunger for western land were British sides to the American character quite alien to France.[100]

Thus, surveying seeped into the army through the dominant culture that anchored the Republic at large. Topographers maintained the nation's most varied collection of British surveying equipment and, in sharp contrast to the French textbooks at the West Point Library, four-fifths of Roberdeau's holdings were works published in English, many from London.[101] Britain remained the preferred destination for first the generation of topographical officers who worked on internal improvements. In 1828, while the Corps was paying for a bright professor to study engineering in France, two topographers were in London to witness the English alternative to French canalization—railroads powered by steam.

Given this cultural inclination, it is hardly surprising that surveyors like Hutchins and Long would share some common assumptions about the strategic importance of frontier navigation. Soldiers, after all, ventured west by following water. Their maps supported the British contention that, because settlers needed access to forests and rivers, the interior was probably uninhabitable and certainly indefensible beyond the major branches of the Mississippi system. Hutchins had neglected the lesser rivers. Soon after the Revolution he had calculated the total expanse of the United States covered by water—fifty-one million acres, an underestimate of 16 percent.[102] Like Samuel Holland, the Americans seemed to believe that rivers were hardly rivers if they were too rocky for commerce and that the interior was worth little if it could not be crossed in a boat. Long also used water to measure the land. Where the topographer had hope for canals—in desolate Illinois, for example—Long reported "a luxuriant growth of blue grass" ideal for grazing.[103] Where the prairie was landlocked, where his expedition took off on horseback after its steamer had choked in the mud, Long encountered the naked edge of his civilization, a blistering waste.

In these ways the art of surveying remained an interpretive tool—reflective, often political, and driven by some of the dogma that guided natural science. No survey is a photocopy of nature. Even photographs taken from an airplane can vary according to light, angle, type of film, and choice of subject.[104] Likewise in the world of Hutchins and Long the strategic mapping of rivers was a task for the mind as well as the eyes. Colored by training and culture, preconditioned by the topographer's need to establish a niche for his expertise, the army's approach to surveying split into double focus—half British, half French. Britain turned reconnaissance west to interior navigation. But French engineering, a West Point tradition, returned surveying to coastal waters where the study of rivers and harbors seemed vital to the common defense.

The West Point Connection

Reporting to the War Department in 1842, U.S. topographical engineer George W. Hughes, a year back from a tour of Europe, chided the British mechanics who dammed streams to feed the canals. Although field-trained builders preferred these kinds of improvements, the remarkable French engineers were clearing and rebuilding the rivers so that channels worked like canals. The Ohio and Mississippi seemed ripe for French innovation, but first, Hughes insisted, Americans needed "science" and "an intimate knowledge of the habitudes of the river." If builders would make the effort to understand stream behavior, then water, indeed "all nature," would dance to the "genius of man."[1]

The engineering institution that nurtured these prospects for science was the army's Athens and Sparta—a center for useful knowledge, a school for professional soldiers, a source of rational planning in American public works. West Point wove the study of navigation into construction at large. From the time of Jefferson to the Civil War, the school promoted a vigorous exchange of waterway information, leading the building professions through three phases of growth: first, when the Corps hired Europeans to teach math and science; second, when graduates studied abroad; third, when innovation at home forced Europe to take note of West Point engineering. But full technological independence, a fourth phase of professional growth, eluded the early West Pointers. Even after the Erie Canal, when Europeans praised the inventive Yankee as a symbol of republican progress, the academy's cosmopolitan science was the bunker of a bookish tradition that distanced army construction from improvised frontier technique.[2]

West Point, after all, had grown from a French conceit. Under the plan tossed about during the American Revolution, the science of Monge and Vauban would teach the minuteman army to outwit a superior force. After 1792 the wars of the French Revolution revived the academy concept. Washington, fearing invasion, launched a fort-building program with seven "temporary" engineers, all Frenchmen. The Federalists also proposed an artillery-engineering school, and on May 9, 1794, Congress created a grade in the army for an officer in training, a "cadet." The same legislation authorized a Corps of Artillerists and Engineers with 764 men in four battalions. It was the first use of the word "engineer" in federal law.[3]

Fortress West Point with its elaborate river defenses seemed like an excellent place to train engineers, but cadets resented formal instruction, and the fear of the French shattered the brittle command. In 1797 a classroom was closed because of a fire. Meanwhile the post commander, Lt. Col. Commandant Stephen Rochefontaine, was torn between a fort project in New Hampshire and his unruly school at West Point. One night the French engineer confronted an angry lieutenant. Rochefontaine struck the man with the hilt of a sword, fought a duel with pistols, declined another, and was vilified in the local press before a board of inquiry demanded his resignation in 1798. At the commandant's trial, a judge admitted that "political considerations" would soon force the French from the army.[4] Meanwhile the Adams administration was looking for help from Prussia. "The knowledge of certain arts and sciences is absolutely necessary," said James McHenry, secretary of war.[5] Inspector General Alexander Hamilton fully concurred. Calling the facilities at West Point "very inadequate," Hamilton wanted no less than five academies—a school for the artillerists and engineers; schools for the cavalry, the infantry, and the navy; and a two-year "fundamental school" to introduce science and math.[6]

Congress settled on just one academy in the Military Peace Establishment Act of March 16, 1802. Conceived by Jefferson and his war secretary, Henry Dearborn, the act, vaguely worded, said "a corps of engineers . . . shall constitute a military academy."[7] The Corps and West Point seemed to be one and the same. What the act left unsaid spoke volumes about the Jeffersonian ambivalence toward peacetime armies. There was nothing about the status of engineer troops or the role of the Corps in battle. Not until 1812 was it clear that West Pointers could serve throughout the army. Not until 1866 was it clear that a nonengineer could command the school at West Point.[8]

A delicate shift in Jefferson's thinking allowed the dove opposition to support the Corps and West Point. Jefferson, a fiscal conservative, was also a states'-rights purist who conceded the need for an army but feared its entrenched command. A school would tame these professional soldiers. Arming builders with science, it would groom the future elite of a loyal Republican army. West Point, moreover, fit bipartisan plans for a national university. Federalists proposed a college devoted to warfare, a University of Mars. Jefferson stressed civilian science. He stripped a bit of the army from engineering by detaching the Corps from artillery units. He struck a body blow at the military establishment by cutting a fourth of its troops. Where others had called for a war college in the name of science, Jefferson did the reverse: he disguised a commitment to science as a minor concession to war.[9]

Another clue to Jefferson's mind was his choice for West Point superintendent: Chief Engineer Jonathan Williams. A merchant, a scholar, a soldier never tested in battle, Williams was an armchair engineer without training in a practical craft. Yet the founder of West Point was a remarkable figure, a true patron of practical science. During his decade at the school, 1802 to 1812, Williams focused the Corps with academic projects that grounded construction in theory and kept French contacts alive.

Williams began his engineering career in the shadow of Benjamin Franklin, his mother's uncle. In 1774 the two men shared a house in London,

Jonathan Williams, chief engineer and first superintendent of the U.S. Military Academy, 1802–1812. Courtesy of the Office of History, U.S. Army Corps of Engineers.

and during the American Revolution they worked together in France. A Harvard graduate, Williams had taken courses in science. In 1785 he helped Franklin chart the Gulf Stream, branching out on his own with a book on navigation and a process for whitening sugar. Plump and stocky with gentle features—he even looked liked Franklin. Williams was also close in spirit and mind to the Sage of Monticello. Jefferson, a cartographer, took an interest in Williams's attempt to measure altitude with barometers. Exchanging notes in 1796, Jefferson and Williams agreed that the Blue Ridge Mountains were the tallest ridge in Virginia. Both scientists became high officials in the American Philosophical Society, and in 1799 the future president joined Williams on a committee to promote the neglected study of Indian burial mounds.[10]

Not until 1801 at age fifty-five did Williams enlist as a major in the short-lived Corps of Artillerists and Engineers. John Adams had searched for a more seasoned builder, someone to rid the army of the disagreeable French. But Williams, although Boston-born, was Gallic in many respects. As Franklin's assistant in France from 1776 to 1785, he had studied some physics, and there he began what became the first American text on harbor fortifications, a translation project. France remained his inspiration. His most famous work as a builder, Castle Williams in New York Harbor, was a tribute to "the greatest Genius in the Science of Defense that the World ever saw": the French engineer Marc René, marquis de Montalembert.[11] One of Williams' first acts as West Point superintendent was to petition for a French instructor. Cadets learned to value the French approach as a blend of disparate fields. "Science," Williams explained, "is in its own nature so diffuse that it is almost impossible to designate the dividing lines." Chemistry and mineralogy converged. Astronomy, geography, physics, mathematics, the sciences of fortifications and artillery fire—these subjects, Williams insisted, were "inseparable" and "interwoven."[12] The branches of army science grew from a common trunk.

The contention that soldiering required science, coming as it did from a neophyte engineer, unsettled those in the army who still feared sedition from France. On July 29, 1802, the War Office made it clear that "in no instance" would the academy superintendent have command over combat forces, even the troops at West Point.[13] But Williams looked abroad and clung to a broader mission. West Point was his "leading Star, . . . a great National establishment to turn out Characters which in the course of time shall equal any in Europe."[14] One day his young cadets would be true men of science, not cloistered technicians but a vigorous corps du génie, America's fighting savants.

1. Science and the Rickety Child

Four months before Williams took charge, the last of Washington's Frenchmen, Maj. Louis de Tousard, resigned from Jefferson's army. Tousard, a gifted scholar, was exactly the kind of Enlightenment soldier that Williams most admired. As an artillerist at West Point he had promoted "le système Gribeauval" that called for firearm standardization through interchangeable parts. Once an aide to Lafayette, the major had lost an arm in a raid on a British position during the War of Independence. He now lost his job and even his pension in the battle for control of the Corps. Although Tousard had survived hysterical times by keeping his distance from President Adams, a new hatchet at the War Department, General Dearborn, effectively purged the Corps. Dearborn captured the mood in a comment he made to Latrobe: the army "had no occasion for engineers," and the general "would never consent to employ foreigners, especially not Frenchmen."[15]

The policy depleted the army. In July 1802, the all-American Corps had a lieutenant colonel (Williams), a major (Decius Wadsworth), two captains, four lieutenants, and nine young cadets. A library begun by Tousard had about fifty books and some charts. Ten years passed before the academy even tried to teach engineering in a single, wide-ranging course.[16]

That the academy survived at all was a tribute to the one field of army science where Americans truly excelled. Reconnaissance, the strategic mapping of water and land, was the nexus of early West Point, the point of common interest where dissimilar backgrounds converged. The school attracted some leading names in geographical science: Williams, Jared Mansfield, Ferdinand Hassler, Andrew Ellicott, Stephen Long, Isaac Roberdeau, Charles Davies—all accomplished surveyors. Captain Mansfield, a Yale graduate, lectured on math and physics while serving on and off as U.S. surveyor general of the Northwest Territory. In 1801 he became perhaps the first native-born American to publish a significant work of applied mathematics, *Essays, Mathematical and Physical*. Jefferson, much impressed, sent Mansfield to West Point where the mathematician began his twenty-five years as one of the army's preeminent scholars. Mansfield headed an industrious clan of army technicians. Nephew Joseph Totten (USMA 1805) rose to international fame as a fort architect and a powerful chief engineer. Mansfield's son-in-law Charles Davies (USMA 1815) was the author of a surveying text and one of America's foremost translators of French mathematics. Another son-in-law, artillerist John O'Connor, translated Baron Simon Gay de Vernon's important *Treatise on the Science of War and Fortifications* (1817).[17]

Surveying was also the passion of Ferdinand Rudolph Hassler, a Swiss with a crowded mind, a true eccentric. Born in Bern and schooled in Paris, Hassler was a book and instrument collector who, as a visiting student at the Ecole polytechnique, had boldly introduced himself to great Enlightenment figures like Borda and Lavoisier. In 1805 he came to the United States as an investor in a utopian agricultural colony, but the venture quickly failed. Searching for employment, he wrote Jefferson and sold some weights and metric rods to the American Philosophical Society. Soon Hassler had a driving ambition: he would triangulate thousands of miles of American coastline, fixing the location of every lighthouse while mapping inlets and harbors. Congress approved the plan in 1807, and Hassler was made "superintendent" of the U.S. Coast Survey, a new maritime bureau in Albert Gallatin's Treasury Department. Hassler spent the next three years teaching math at West Point and searching for equipment and funds. After the War of 1812 at last he began the project with help and some resistance from topographers William G. McNeill and J. J. Abert. Here began a close, often competitive relationship between the government's two active centers of water reconnaissance—the U.S. Topographical Bureau and the U.S. Coast Survey.[18]

Hassler's departure in 1810 left the West Point math department to Capt. Alden Partridge, a former cadet. Partridge loved to drill, and he led surveys through the nearby mountains, testing equipment. When Partridge became West Point superintendent in 1813, the math chair passed to one of Jefferson's favorite surveyors, Andrew Ellicott. "In Europe," wrote Ellicott of his appointment, "the first scientific characters are attached to their military academies, and there, as well as in this country, the professor of mathematics is considered the principal or president of the institution."[19] Principal or not, Ellicott became the intellectual light of the academy during the dark years of war and shifting assignments, 1813 to 1816.

Even in Ellicott's time the school for engineers still had no regular program of prerequisites that took students from theory to practice. Instead the army made do with drafting, drawing, lectures on tactics, a math unit on "trigonometric surveying," and an astronomy lesson on "different methods of determining geographical points."[20] West Point also had an unusual science cabinet: instructional models, drafting instruments, a thirty-inch refracting telescope, theodolites, and globes. The library's strength was surveying, astronomy, and general science. About a third of the titles were French.[21]

Although the library expanded rapidly, reaching 942 titles by 1822, good textbooks were scarce. At first the school relied on British texts, such as Charles Hutton's *Course of Mathematics* (1801) and William Enfield's *Institutes of Natural Philosophy* (1802). Both applied mathematics to a range of practical problems—carpentry, surveying, hydraulics. Cadets learned to determine "the strongest angle of position of a pair of gates for the Lock of a canal or river," and from that position to calculate "the quantity of pressure sustained by a dam or sluice." After the introduction to hydrodynamics came a problem in lock construction on Hutton's final exam:

> Suppose the breadth of a canal at the top, or surface of the water, to be 24 feet, but at the bottom only 16 feet, the depth of water being 6 feet. [Solve for] the pressure on a gate which, standing across the canal, dams it up.
>
> Answer: 9 tons, 80 pounds.[22]

Elementary, perhaps, and not yet engineering, but Hutton and Enfield anticipated the future of the West Point academy by relating hydraulics to surveying and transport planning. Textbooks were becoming a thread of technology transfer that linked building to elite academe.[23]

Meanwhile the pathbreaking work of Jonathan Williams kept the focus on currents and tides. Williams, watching Benjamin Franklin, had begun to investigate water on long trips across the Atlantic. One inky night a storm blew his ship against a rocky coastline. Badly shaken, Williams began sounding the ocean, searching for ways to detect reefs and shoals. In 1793 he published an original map and later a "memoir" called *Thermometrical Navigation* (1799). The memoir rejected the notion that fires raged under the sea and heated the ocean floor. Rather, said Williams, the depths of the ocean were ice. Shallow water was colder than the water of the mid-Atlantic that mixed with tropical currents. Although Williams admitted that "the observations of a mariner are more likely to be attended to by mariners than any instruction given by landsmen," hydrographers were impressed.[24] Ciprino Vimercati, director of the Maritime Academies of Spain, called the temperature experiments "minute," "progressive," and "analogous to . . . physical principles, universally admitted, concerning the continued action of heat and distribution of it through all bodies."[25] Williams, already known for clever work with telescopes and barometers, now pioneered a practical use for thermometers at sea.[26]

Reprinted in London, Madrid, and St. Petersburg, *Thermometrical Navigation* gave the infant Corps some instant stature, and it previewed the future of West Point engineering by touching on central themes: protracted experimentation, the study of stream behavior, and the Franklin-like promotion of navigation surveying as engineering science. Water reconnaissance

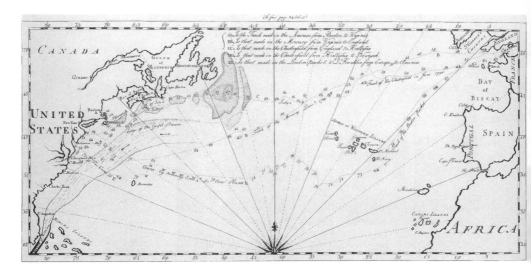

Map of the Atlantic Gulf Stream with ocean temperatures from "Memoir of Jonathan Williams" (1793), reprinted in *Thermometrical Navigation* (N.p.: 1799). Courtesy of the American Philosophical Society.

expanded the army's approach to engineering education. Although historian Theodore J. Crackel has minimized the importance of scientific instruction at early West Point, claiming that the textbooks were remedial and that men like Hassler and Mansfield were "laymen" rather than "professional scientists," Williams kept scientific ideals alive.[27] He built instructional models. He sponsored translation projects and contributed at least fifty books and pamphlets from his own technical library. He also pushed engineering as science through the United States Military Philosophical Society, an outlet for Corps research.

Chartered on November 23, 1802, the Corps' philosophical society promised to "preserve as far as possible the military science which must still exist in the different states." It hoped to serve as an army forum open to all citizens and "exclusively confined to the Arts and Sciences"—in short, hardly exclusive at all.[28] Williams presided as "president." Jefferson was listed as "patron." In 1805, after three years of confusion while Williams feuded with Dearborn, the society at last lived up to its charter, holding meetings about twice a month at West Point. Five years later the society had 216 members—past and future presidents John Adams, John Quincy Adams, William H. Harrison, Madison, and Monroe; five governors, eight senators, a mayor of New York, a Supreme Court justice, inventors like Robert Fulton and Eli Whitney, frontier soldiers like Charles Gratiot and Zebulon Pike.[29]

Given Williams's orientation, it helped to have an interest in construction and water research. Francis Masson, teacher of French, outlined his plan to translate "all that is known in Europe" on fort engineering.[30] Hassler spoke on canals and the coast survey. Although the historical record is spotty, a clear record of one active session shows the West Point taste for applied hydraulics. On October 6, 1806, the society spanned water topics from hydrostatics to harbor defense. Papers included:

An Hydraulic memorandum describing an invention of the celebrated Mr. Montogolfier, on raising by pressure a greater quantity of water.

An original thermometrical journal round Cape Hatteras, proving that shoals may be discovered by comparing the relative heat of the water.

[The] course of all the meanders of French Creek, from Lake Le Boeaf to the Alleghenies river, and thence down river to Pittsburgh.

A chart of the harbour of New York.

A rough draft of the Road from Erie to Waterford containing an actual measurement of the falls of Niagara, . . . Height of bank from river at Niagara, Distance across the strait.

Other papers discussed gunboats and "the principles of marine batteries"; still another, the improvement of life rafts. "Hydraulics," Williams explained, "make a very important branch of Military Science," and the chief engineer hoped his young engineers would "pursue the investigation in other places under other circumstances."[31]

Williams then took the final minutes of the upbeat session to discuss the geology of the rapids at Louisville, the falls of the Ohio. Once, long ago, the river had narrowed and a great raft of logs built a natural dam. Over time, calcium in the water petrified the logs, creating limestone. Williams said nothing about blasting the rocks or canaling around the obstruction. More theory than practice, he defended a higher purpose: "to instruct in the changes that have taken place on the surface of our globe, their causes and effects, . . . especially as knowledge of this kind is often useful in military operations."[32]

Suddenly, in 1807, the army had a pressing need for research on rivers and harbors. A French warship had entered the Delaware River. A British squadron guarded the coast of Virginia, threatening Norfolk. When the royal frigate *Leopard* fired on the USN *Chesapeake,* killing three, outrage swept the seaboard. Jefferson had to respond. Pushing a divided Congress, the president demanded an ironclad trade embargo, and in 1808, Jefferson launched a "second system" of coastal fortifications (the "first system" being the forts begun by Washington's Frenchmen in 1794). Jefferson knew that the same imperfect highways that took Americans west were rivers and natural inlets open to sudden attack. His plan called for land batteries, movable artillery, barges with mortars and cannon, and about two hundred European-style gunboats anchored at the entrance to harbors from Louisiana to Maine. As the regular army nearly tripled, Jefferson put out the call for fifty thousand volunteers. Virginians rushed troops to Norfolk. Congress debated a plan to seal off the port of New York by sinking timber spikes in the harbor.[33]

In the summer of 1807 Chief Engineer Williams seized the chaotic moment to advance a radical plan. Instead of flat, star-shaped defenses that dated from the time of Vauban, the Corps wanted tall, circular forts, a design that originated in France with the "perpendicular" system of engineer Montalembert. Tall forts were like ships walled with cannon. Two or three stories high, their guns bricked into shellproof chambers (casemates) with narrow windows (embrasures), the new American forts would tower over the water, amassing a deadly fire at critical points in the channel where an enemy fleet might try to attack. Engineers raced to acquire an exact sounding and survey of every vulnerable inlet. From 1808 to 1812, Williams and five future chief engineers got their feet wet, literally, on fort-building assignments. Joseph Swift, Joseph Totten, Walker K. Armistead, and Charles Gratiot all surveyed the seaboard, planning defenses. Alexander Macomb proposed a canal through a tiny island at Georgetown, South Carolina. Williams began a castlelike circular fort at Governors Island in New York Bay. By 1812 the Corps and the War Department had spent about three million dollars on forts at thirty-one sites. These were the first harbor projects tied into a national system, planned by West Pointers, and built almost exclusively by native-born engineers.[34]

Circular forts made a lasting impression. Although the Corps completed only a few before 1812, casemated gun chambers with slitlike windows would dominate the "third system" of coastal forts built in Jacksonian times. Elaborate plans for these kinds of defenses revealed more about principal harbors than the next thirty years of plodding triangulation by Hassler's coast survey. Fort building, moreover, advanced army hydraulics in more immediate ways. Shoreline beach construction sent engineers searching for ways to combat erosion and stabilize the ground. Looking chiefly to Holland and France, the Corps learned about seawalls, breakwaters, fascine mats that retarded erosion, and waterproof, lime-based cement. Fort building also made

Castle Williams at Governors Island, New York. Designed by Jonathan Williams and completed in 1810, the three-tiered circular fort was based on a French design. Courtesy of the Special Collections, U.S. Military Academy.

Williams's point about the need for science training. Circular forts were proof that theory could alter design.

Still, at the close of its first decade the Corps at West Point remained, as Williams put it, "a puny, rickety child."[35] Neglected by the War Department and cut off from Congress, the cadets in patches and rags were "like Falstaff soldiers . . . as bare as newly shorn sheep," but the swelling demand for native builders was a chance for rapid promotion.[36] Swift became a lieutenant colonel after ten years in the army. Macomb, rising swiftly, was a brigadier general at age thirty-two. Their status bred deep resentment among officers of the line. In July 1812, less than a month after Congress declared war on Britain, eighteen junior officers shot off a letter of protest that blocked the chief engineer from a combat command. Williams promptly resigned. His sudden departure effectively beheaded the Corps, West Point, and its forum for science research. Only six cadets and one professor showed up for the fall semester. In December the fuel ran low and classes were suspended. The U.S. Military Philosophical Society laid dormant for about a year until a few members, meeting at Washington Hall in New York City, voted at last to disband.[37]

There was one stubborn voice of dissent at that last recorded meeting of the society during the War of 1812. Future West Point superintendent Sylvanus Thayer, voting alone to revive the organization, continued to study construction and readied himself for the day when the tiny school on the

Hudson would become what its founder intended—a leading star of math and science, an American polytechnique.

2. A Lesson in Hydraulic Construction

"Perhaps without exaggeration," wrote historian Henry Adams, "the West Point Academy might be said to have decided, next to the navy, the result of the war." [38] Although the war was a bloody stalemate, its military result unclear, the former cadets did well enough to vindicate the academy concept. In 1815 the War Department again made the case for science instruction. A bill to build more West Points died by a single vote. [39]

The War of 1812 also exploded the myth of the gentleman builder who was too much the scholar to fight. Lt. Alexander J. Williams, Jonathan's eldest son, was the only West Pointer actually cut down in combat, but others served with distinction. A small force under Macomb's command fought back a wave of British at Plattsburg, New York. Swift, cool under fire at the Battle of Chrystler's Field, was promoted to brigadier general and took charge of the port of New York. His twenty-six engineers had fortified the

Sylvanus Thayer, "Father of the Military Academy" and its fourth superintendent, 1817–1833. Courtesy of the Special Collections, U.S. Military Academy.

army, housed it, moved it, mapped its advance and retreat. Their defenses had held at Lakes Erie and Ontario, at Baltimore and Norfolk—in fact, no fortress built by the Corps had fallen to enemy fire. While this had as much to do with location as strength, Congress commended the Corps, and the war left the widespread impression that the army had profited from technical education. The stage was set for reform under a new West Point superintendent and a new secretary of war.[40]

Superintendent Thayer and Secretary Calhoun were, in manner and style, a black-and-white study of contrast, yet in devotion to French engineering the two were of one mind. Straitlaced and aloof, Thayer spoke with a measured calm that hid great passion for France. Classmates at Dartmouth remembered the meticulous Thayer as an avid reader of a Washington paper that followed Napoléon's every campaign. Later as a cadet and junior officer based at West Point, Thayer's quiet ambition was to see his hero in action. His chance came when Napoléon escaped from the island of Elba in February 1815. Thayer requested a furlough, but Brigadier General Swift, eager for contact with Europe, sent the young man abroad on active duty with engineer William McRee. The trip was a crucial event in American engineering. Although it was shocking to see France as a broken nation after Napoleon's final defeat, Thayer and McRee spent almost two years in the company of Gen. Winfield Scott and others who were studying warfare in Paris. They combed libraries, inspected forts, tested instruments, attended classes and met professors at the Ecole polytechnique, and visited the advanced school at Metz. They also purchased charts and seven boxes of books for West Point, more than a thousand items.[41] At least eighty titles addressed some aspect of transport planning. Returning to West Point in July 1817, Brevet Major Thayer, now superintendent at age thirty-two, launched a grand program of academic expansion *à la mode française*.[42]

Thayer was about two months into the fall semester before President Monroe found a brash insider to expedite army reform. Calhoun of South Carolina, just three years older than Thayer, had risen to national stature after six fiery years in the House. A bellicose spokesman for the Republican War Hawks, Calhoun did for the army what Hamilton and Gallatin had strived to achieve at the Treasury Department: he virtually abolished a massive debt, and with a handpicked general staff he reorganized the department so a leaner, more efficient army could rapidly expand. Thayer fit into these plans. Both men applauded the military rigidity of the French educational system, and Calhoun, an acquaintance recalled, was "almost fanatic in . . . bias for & admiration of French eng[ineering] and French soldiering."[43] If Calhoun's general staff was to serve as the rational head of a skeleton army, a beast that

John C. Calhoun, secretary of war, 1817–1825. Courtesy of the Office of History, U.S. Army Corps of Engineers.

could grow to contain a French-style levy en masse, then Americans needed a top-heavy corps of commanders with superior qualifications—tacticians and professional soldiers, men taught to lead raw recruits.[44]

Calhoun's lasting success was a step toward professionalization known as the "Thayer system." Adapted from the *grandes écoles* at Paris and Metz, it was martial law in the classroom, a Spartan regimen of rote and recitation: daily reports, weekly testing, semiannual comprehensive examinations, a uniform nine-month curriculum with four academic years and a five-year minimum enlistment. It was also a method of ranking cadets by competitive performance, a "merit role." Drill, sobriety, almost monastic isolation—the Thayer system was a science boot camp, a hair shirt for the sons of privilege. Andrew Jackson thought it was autocratic. "The very thing taught," said a former cadet, was "*positive obedience.*"[45]

The choice for professor of civil engineering confirmed the commitment to France. He was a graduate of the Ecole polytechnique and the two-year artillery program at Metz, a tall, heavyset man with deep, intelligent eyes—Capt. Claudius Crozet. A soldier with a colorful past, he had been a bridge builder, a *pontonnier,* in the imperial artillery corps attached to Napoléon's headquarters during the Battle of Wagram (1809). As an engineer in Holland and Germany he studied the sluices and navigation jetties that were later described at West Point. Just how he came to New York has not been recorded. One biographer says the talented captain sailed into exile with Napoléon's aide-de-camp, the fort builder Simon Bernard.[46] Perhaps it was Bernard who recommended Crozet to West Point. Arriving with his Parisian bride in September 1816, Crozet joined the three other Frenchmen already on staff. Four out of seven West Point professors were now foreign-born.[47]

Science professor Jared Mansfield thought the Frenchmen were an annoyance, but his son, cadet Edward Deering Mansfield, valued their tortured attempts to introduce foreign ideas. Crozet, especially, was a man of "inexhaustible patience" who labored to teach engineering but found the cadets unprepared.[48] Discouraged after a year and still struggling with English, Crozet returned to the basics. He taught geometry from French engineering diagrams, and in 1817, he began using a blackboard to project three-dimensional figures on a coordinate plane. He also assembled an important text, his *Treatise on Descriptive Geometry* (1822). Neither the math nor the method of instruction were new to West Point. The *Treatise* drew from books by Blaise Pascal and Gaspard Monge. Some of the same geometry had been introduced by Hassler in 1808, although less systematically. Crozet vowed only to "follow the methods of instruction and the authors of Ecole polytechnique, especially those lessons which I myself have received and which I have employed."[49] Yet, as an agent of this transfer, the Frenchman deserves credit for formalizing the order of educational prerequisites that paralleled the intellectual development of the polytechnic engineers. Before they could build, cadets learned to draw, survey, and project a foundation mathematically from diagrams to the ground.[50]

Crozet's choice for an engineering textbook, *Program D'un Course de Construction,* took the same French approach. Compiled from the lecture notes of Joseph-Marie Sganzin, the road and canal expert at the Ecole polytechnique, the text stressed the need for elaborate planning. The section on seaports was a lesson in mapping and sounding. A chapter on "Works of Navigation" taught builders to ponder their options: would they build a slack-water project, dredge sandbars, blast rocks, or avoid these hazards altogether with a dry-land canal? The choice rested to a large degree on the sharpness of grade. If the incline was steep, a builder might try pooling the river with navigational dams. If, however, the slope was less than three feet per thousand, a sandy channel could be dredged, and an engineer might facilitate navigation by removing rocks and building a towpath. In all cases the builder began by sketching a profile, a chart, or even "a complete representation of the bottom . . . similar to the representation of a hill or mountain"—in essence, an underwater survey.[51]

Sganzin succeeded where others had failed by relating civil construction to the science of measuring land. In 1826 the book's translator explained that "nearly all" construction manuals were "too voluminous" or "too much involved in mathematical language to be accessible or intelligible to the greater part of learned and practical mechanics."[52] Sganzin's *Program D'un Course de Construction* appealed to a younger and broader audience, yet the book was Continental in scope. It took readers from Flanders to Languedoc with the kind of exotic examples that West Pointers would later use to justify army projects. Later English-language editions balanced the lesson on navigation with a few words on British and American achievements, but the treatment of rivers remained novel and decidedly Continental—probably the first textbook in America to apply trigonometry to dam construction, the first to discuss seaports, lock design, reservoirs, and slack-water canalization.[53]

Not until 1830 did the New Yorker Amos Eaton provide an alternative to Sganzin with a more basic text, *Art without Science*. Here the subject was Americanized with local examples "divested," said the subtitle, "of the speculative principles and technical language of mathematics."[54] Since 1824 the geographer Eaton had worked with Stephen Van Rensselaer to promote civilian engineering at a school in Troy, New York. By 1838 only seven graduates of the Rensselaer school had become engineers. Meanwhile, more than one hundred West Pointers had joined the civilian profession.[55]

West Point had other rivals but none that could equal the army's reputation as a source of builders, educators, and transportation experts. Reporting on the school's engineering course in December 1834, an editor of the *American Quarterly Review* found it "a comprehensive and liberal one, and [an education that] reaps the full benefits of all the previous preparations which the cadet has received in the study of mathematics and natural philosophies." Moreover, the article continued, the new engineering professor had enjoyed the "rare privilege of studying these subjects in the military academies of France," and had imported from Europe "all the latest improvements."[56] Here in simple language were lasting characteristics of the West Point approach: mathematics, a pioneering interest in scientific theory and method, and unshaken admiration for France.[57]

These traits found high expression in the work of a frail professor who for forty-one years was, next to Thayer perhaps, the most feared man at West Point. Dennis Hart Mahan, the son of an Irish immigrant, had a cutting, sarcastic manner that stunned and dissected cadets. A former cadet himself, the young Mahan had cared little for the pomp of the army, but at age seventeen he had decided to look into West Point because the school taught drafting and Mahan loved to draw. His freshman year, 1820, was a trial period for academic reform under the Thayer system. Crozet still taught engineering (although he left when Mahan was a junior), and twelve of the twenty-six introductory texts were still available only in French. First- and second-year

Dennis Hart Mahan, West Point professor of civil and military engineering, 1832–1871. Courtesy of the Office of History, U.S. Army Corps of Engineers.

mathematics—six hours a day, six days a week—relied on lessons prepared by polytechnic scholars like Carnot and Monge. Fourth-year engineering introduced seven subjects: "elementary parts of buildings and their combination; orders of architecture; construction of buildings and arches; canals; bridges and other public works; machines used in construction; execution of a series of drawings, consisting of plans, elevations and sections."[58] Mahan excelled. At graduation in 1824 he was Thayer's top scholar with a remarkable score of 19.38 of 20.00 possible points.[59]

Mahan, for much of his life, suffered from lung infections, but for two years he held on at West Point as a lieutenant and assistant professor. In April 1826, his health deteriorating, Mahan asked for leave to a warmer climate, preferably France. Here began a West Point tradition of semiofficial travel to Europe on the pretext of health.[60] Mahan left in July with support from Congress and a $500 advance. Paris in 1826 was still the Mecca of science—the center, said Mahan, of "all that is distinguished in Europe for talent."[61] Mahan made a careful study of drawbridges, gun emplacements, construction machinery, wood preservatives, and new kinds of French cement. He toured the grand canals, sketching locks. Unable to attend the famous bridge and highway school, he settled for the accelerated engineering and artillery courses at Metz, a two-year program. Upon his return to New York, in July 1830, Engineer Lieutenant Mahan was probably the most educated man in the Corps.[62]

In 1832 it surprised no one at West Point that Thayer favored Mahan for Crozet's old professorship in the engineering department, a job he held until his death in 1871. At first he taught construction from his own collection of sketches. The problem, according to the *American Quarterly Review,* was the "want of a sound textbook"; for there existed no modern work in English, none that discussed hydrology, materials testing, or truss-bridge design.[63] Mahan also recognized that the academy's introductory text, now about thirty years old, was "almost a dead letter to persons who are entirely unacquainted with the subject of construction."[64] Compiling his sketches and notes in 1837, Mahan published the most widely read text of its kind, *An Elementary Course of Civil Engineering, for the Use of the Cadets of the United States Military Academy.* The book was a large success. Reprinted in London and elsewhere in Europe, *Civil Engineering* sold more than fifteen thousand copies from India and China to technical academies such as Rensselaer, Norwich, the Virginia Military Institute, and The Citadel Military College of South Carolina. Mahan's final revision went through twelve editions and survived as a standard reference in the West Point library until the first decade of the twentieth century.[65]

Despite its popularity in civilian schools, *Civil Engineering* remained military in its distinctive mix of mathematics, science, and engineering theory. Mahan considered quantification "indispensable to the successful pursuit of this profession," and he argued that "from a proficiency in mathematics one can nearly always foresee the proficiency in engineering."[66] "Science," in Mahan's vocabulary, was the use of sophisticated instruments and complex surveying techniques. Science was also theory—the principles derived from field experimentation, the laws that reduced construction to standard procedures. It was the physics of dam construction, the hydrology of canals, the study of waves as applied to breakwater design, the static theory of arches. "My boy, remember one thing," Mahan told his son, "the only really practical man is the one who is thoroughly grounded in his theory."[67]

In America this foundation in theory placed Mahan in the minority camp, for in 1837 the great majority of chief and supervising civil engineers—79 percent according to Daniel Calhoun's data—were field-trained craft engineers. The first edition of Mahan's *Civil Engineering* virtually ignored frontier construction by highlighting the achievements of "the most scientific corps in the world, the French Corps of Engineering and artillery."[68] The lesson on fluid mechanics, for example, was based on an 1835 review by a French-educated Bostonian, the dam builder Charles S. Storrow. The analysis of hydraulic cement was taken from studies by a French general of engineers, Clement Louis Treussart.[69]

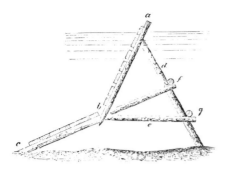

Cross-section of an Italian wing dam, from Dennis H. Mahan, *An Elementary Course of Civil Engineering*, 5th ed. (New York 1851). Courtesy of the Office of History, U.S. Army Corps of Engineers.

Even in river and canal engineering, a burgeoning American field, Mahan cited the Po, Rhine, and Loire, but the first edition of *Civil Engineering* said nothing about the work of the Corps on the Mississippi and Ohio. The book's fifth edition (1851) continued to slight the frontier with French and Italian examples. Mahan referred to the "inconveniences" resulting from narrow, shallow canals, a veiled slap at the Erie system.[70] Mahan also acknowledged French standardization and, like Crozet, had little use for small, temporary construction. Later editions of *Civil Engineering* were kinder to American projects, but the European bias endured. The Frenchmen Poncelet, Treussart, Claude-Louis Navier, and their contemporaries continued to find an American audience in Mahan's more specialized studies, most notably his *Complete Treatise on Field Fortification* (1836) and his textbook on industrial drawing (1852).[71]

Mahan honored these sources with no claim to originality. His ideas were from "the common stock belonging to the profession in general," ideas applicable to many civilian fields.[72] Chief Engineer Swift had already proven the point by resigning from the army to work as a harbor surveyor. Crozet became state engineer of Virginia. Of the cadets in Mahan's own class (1824) 6 of 31 wound up as privately employed civil engineers. The army, in fact, lost about 60 percent of the West Point cadets during the school's first two decades, and during the 1820s the rate of attrition continued to climb. Of the 554 cadets appointed as of September 1819, only 112 were promoted into the army. Ten years later just 18 of the first 100 graduates remained on active duty.[73]

Thayer and Mahan well understood that cadets lost to the army were not necessarily lost to the nation—they entered every scientific field, dominating astronomy and coast surveying, transforming heavy construction. Railroad engineer William McNeill (USMA 1817) designed the massive, still-functioning dry dock for the U.S. Navy in Brooklyn Harbor. His colleague

SECTION ROOM

An exam before an
officer-instructor in a
West Point section room,
about 1840.
Courtesy of the
Special Collections,
U.S. Military Academy.

and brother-in-law George W. Whistler (USMA 1819) brought modern rail technology to czarist Russia. Mahan's students built some of America's biggest dams, largest canal locks, and longest bridges. Engineer Herman Haupt (USMA 1835) wrote a definitive work on the theory of trusses, and in 1856, he resigned from the Pennsylvania Railroad to supervise one of the great construction feats of the nineteenth century—the Hoosac Tunnel. By 1850, according to Edward Mansfield's statistics, 180 West Pointers had become civilian builders. Another 220 were doctors or lawyers, 100 were professors or college presidents, and 80 had been elected or appointed to high public office. Although some said the civilian West Pointers had fleeced the army for a free education, Mansfield, writing in the *American Journal of Education,* had a different interpretation: the academy had imported a refined "scientific culture" that prepared young men for a range of useful pursuits.[74]

In the field of rivers and harbors the West Point "culture" of science remained the polytechnic orientation that promoted theory and standardization at the expense of frontier technique. Williams had made hydrographical mapping a focus of fort engineering. Thayer, Crozet, and Mahan had formalized the investigation, reducing water improvement to basic procedures and laws. It now remained for the students to show the professors that science applied to construction could elevate the Republic through complex public works.

3. The Transatlantic Profession

If West Point was out of step with craft tradition—its theory abstract, its methods at odds with frontier construction—then what role did the engineers play in the factious but thriving profession of surveyors and building mechanics? How did soldiers interact with civilians, and what resulted from the exchange? In 1957 the historian Forest Hill seemed to suggest that the army was common ground between theory and practice, a passage between technical worlds. That interpretation works best for the half-civilian topographical officers who, like McNeill and Long, sought out lucrative private employment as canal and railroad surveyors. But even in the private sector the lines of contact were strained. Captain McNeill, reporting to the Baltimore and Ohio Company in 1830, referred to the "bad taste" and "rustic manner" of common mechanics, and Long used the same report to denounce the imprudent civilians who questioned his bridge designs.[75] Meanwhile the chief of the topographical bureau tried to break the army's reliance on competent but unschooled "citizen" engineers.[76]

Civilians, for their part, often resented the army. "Blind and ever ignorant obedience" was antithetical to what the geodesist Hassler called the "zeal" of civilian science.[77] Engineer Philip E. Thomas, responding to Long's report on the B. & O., claimed that plans pushed by the army were not only inappropriate, they were "exceedingly obnoxious."[78] Canal builders campaigned to prohibit the private employment of topographical engineers, but restrictions were hard to enforce. Reconnaissance blurred soldiering into science. Competition bred dispute. In 1850, bridge engineer Charles Ellet refused to work with the army on the Mississippi Delta Survey. The Smithsonian's Ferdinand V. Hayden thought an army expedition was "the last place to do a large business in Natural History," but in 1857, he agreed to join topographer Gouverneur K. Warren on a study of the northern plains. The two men learned to hate each other. After a summer mapping Nebraska the scientist confessed a desire to shoot his army companion. Warren, Hayden protested, had treated him "like a dog."[79]

Given the gulf of misunderstanding between soldiers and civilians, it is not surprising that the private sector was slow to embrace *la technique*. Still, as agents of technology transfer, the West Pointers were vital to civil construction—well connected and privileged with sensitive information, schooled enough to record what they saw. Some traveled freely with diplomatic papers. Many had a line of government credit to purchase technical books. As the first class of American builders with the institutional support to travel extensively beyond Great Britain, they witnessed a radical change in hydraulic construction, a "scientific revolution." Their earliest contacts were primarily French. Later, chiefly through the work of a reorganized topographical bureau, the engineer-officers experimented with German, Italian, Dutch, and British technologies that exposed the practical Yankees to a strange new world of ideas.

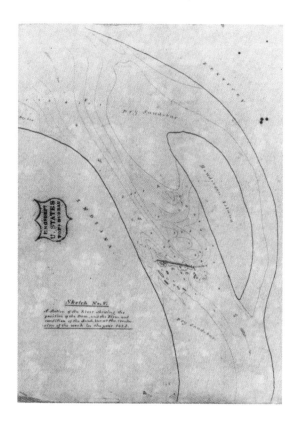

Major Stephen H. Long's
experimental wing dam in
the Ohio River, 1825.
Courtesy of Record Group 77,
Cartographic Division,
National Archives.

One way to disseminate scientific information was through book-buying, fact-finding missions. From 1815 to 1861, the army sponsored or endorsed more than a hundred trips to Europe, most of them to France. West Point–educated fort engineers took at least nineteen excursions abroad before the Civil War. One stayed in England, one studied the war in Crimea, and the rest based their visits in Paris or Metz. Topographical officers stayed longer in England but six out of ten still spent furloughs in France. Although *The Savannah Republican* believed that "nearly every corps of our army is represented abroad," the West Point engineers and artillerists made most of the trips.[80] Corps and topographical engineers took thirty-one excursions to Europe before 1861. West Pointer Alfred Mordecai, an artillerist, took three separate trips. Thayer and Mahan each traveled twice.[81]

In Europe the travelers shopped for equipment, instructional models, drawings, plans, maps, and books. Thayer, anticipating the annexation of Texas, searched for fort models and a French map of Mexico. Mahan imported a small printing machine to reproduce handouts for class. Text-books were shipped to West Point, and by 1861 the academy listed 137 titles on some aspect of hydraulic construction—28 titles from Britain, 80 from France. In the field of port engineering, where Britain had made advances in pier and lighthouse construction, more than half the holdings were French. French canal books outnumbered British titles by a factor of four to one.[82]

Books helped the army rediscover ancient techniques. One mixed success of an ancient technology was the construction of dikes and dams that controlled floods and flushed river deposits, an Italian innovation. Since the time of Galileo and Guglielmini the Italian savants had experimented with structures that enlarged a channel's carrying capacity by increasing the velocity of its current. By the 1770s the French students of Italian hydraulics were forcing sluggish, flood-prone rivers through rows of sunken piles. West Point textbooks referred to these experiments, and in 1821 the engineers Bernard and Totten detailed the operation of low wooden "dykes" that worked by "diminishing the velocity of current above them, thereby economising the expense of water, at the same time constraining the current to rush with greater velocity through the narrow spaces to be deepened."[83] Noting the similarity between sandbars in the Ohio and those in the French Loire, the engineers identified twenty-one promising sites for dike experiments. Major Long built the first experimental structure on a shallow sandbar at Henderson Island in the Ohio. Completed in 1826, he called it a "wing dam"—a double row of piles that slanted into the low-water channel at about forty-five degrees. The dam scoured the bar and was "highly satisfactory" according to the Louisville *Public Advisor.*[84] An accumulation of gravel and sand held the structure in place until the army made repairs in 1872.[85]

The Henderson Island experiment seemed to verify *la technique,* but the men chiefly responsible for western rivers, the topographical officers, gradually lost faith in this kind of construction. Long contrasted the "insignificance of the wing dam" with "the majesty of the Ohio floods." Dams that narrowed a channel could only "give some degree of permanency to the bars, in as much as they will prevent the current from removing the sand from the upper slope of the bar and depositing it on the lower."[86] The topographer Hartman Bache also questioned the reliance on structures that deepened a river at one point but blocked it at another. Bache joined Long, J. J. Abert, and others in the topographical bureau as advocates of steam-powered dredging, underwater blasting, and other cost-effective alternatives to wing-dam construction. In this way, an Italian idea, borrowed largely from French experience, had been tested on the frontier and then set aside for more expedient solutions to hazardous navigation.[87]

The army had more success with seawalls and sunken foundations held together with a kind of concrete the French called *béton*. Rubble concrete of various kinds had been studied by Europeans since the Romans made a mortar from limestone in about 200 B.C. John Smeaton improved the Roman mixture by adding English clay, and the New Yorker Canvas White, recently returned from England in 1818, discovered an American clay that bonded underwater, a native cement. While canal building opened an American market for brand name hydraulic cements—Portland cement from England, Rosendale cement from New York—quality varied. Engineers still searched for a cheap, reliable compound that would withstand crashing surf. Mahan, reporting from Paris, followed French experiments with a *béton* mixture of burnt limestone and volcanic sand. Crushed into powder, limestone was burnt in a kiln, mixed into paste with water and gravel, and used as a mortar or cast into heavy blocks for breakwater construction. Chief Engineer Totten, a talented chemist, found three kinds of American limestone suitable for *béton*. After twelve years of tests at Fort Adams in Newport, Rhode Island, Totten translated a French treatise on mortar chemistry, and in 1838 he introduced the Franklin Institute to American concrete in *Essays on Hydraulic and Common Mortars and on Lime-Burning*. A year later the Corps used a concrete *béton* in an experimental seawall at Oswego, New York. West Pointer Quincy A. Gillmore featured lime burning and the French science of mortar testing in two important handbooks—*Practical Treatise on Limes, Hydraulic Cements, and Mortars* (1863), and *Practical Treatise on Coignet-Béton* (1871).[88]

Although there is no early record of the Corps using blocks of concrete in place of natural stone, a practice pioneered by the French in 1833, engineers did use *béton* in walls, dams, and beach foundations relentlessly pounded by surf. In Boston harbor, for example, where shoaling at Lovell's Island threatened to shut down the port, Colonel Thayer protected the beach with a row of granite boulders set in a trench of *béton*. At fort projects in Boston, New York, and Key West the engineers poured a cement-concrete mixture over broken stones. Corps engineer John Sanders invented a mixing machine. Yet *béton*, like navigation dams, remained an experimental technology tested by the army but seldom used in river construction until after the Civil War. Not until 1878 did the Corps use *béton* as a mortar in a dam foundation, and the era of French domination had passed before the army found a way to build an entire lock chamber out of concrete near Hartford, Kentucky, in 1895.[89]

French theory gave the American army the theoretical justification for truly massive construction. Army piers, dikes, and breakwaters profited directly from the French science of tides. French theorists such as Pierre-Simon de Laplace and Amand Rose Emy had argued that an undertow

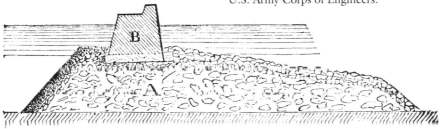

Cross-section of the Cherbourg
Breakwater, the model for the federal
breakwater in Delaware Bay; from
Dennis H. Mahan, *An Elementary Course
of Civil Engineering*, 5th ed. (New York
1851). Courtesy of the Office of History,
U.S. Army Corps of Engineers.

"ground wave," rolling along the floor of the ocean, propelled the powerful
swells that broke violently in the surf. Engineers hoped to disrupt the surf
with sloping walls that cut into the ground wave. At Delaware Bay, for example, the Frenchman Bernard planned a wedgelike breakwater with a pier
to block floating ice. Begun in 1829 and built slowly over the next seventy-
three years, the breakwater never lived up to the Corps' high expectations,
but West Pointers remained committed to French engineering. Bernard and
Totten went on to design a French-style tidal lock and "tumbling dam" to
maintain deep water at the Georgetown terminus of the Chesapeake and
Ohio Canal. Soon after the Civil War the French connection inspired the
Ohio River canalization project and the giant river lock at Davis Island
below Pittsburgh, then the world's largest movable dam.[90]

Meanwhile the army elite so revered French engineering that British
bombardment techniques and Prussian educational reform came to America
slowly and chiefly through Paris. In 1840 the chief engineer made enemies in
Congress with his stubborn defense of grandiose "third system" fortifications planned by Simon Bernard. Mahan, undeterred, organized a Napoleon
Club to study French tactics. Captain Mordecai, a student of Mahan's returning from a second tour of France in 1841, founded a native strength-of-
materials science with impressive experiments on gunpowder and cannon.[91]

West Pointers were by no means the only Americans to study in Europe,
but the few civilian converts to French engineering—builders like Charles
Ellet and Charles Storrow—faced some of the same suspicion typically reserved for the Corps. Ellet and Storrow went to France at a time when many
building mechanics denounced applied science and math. "It is sometimes
objected to the study of mathematics that it contracts the mind," said the

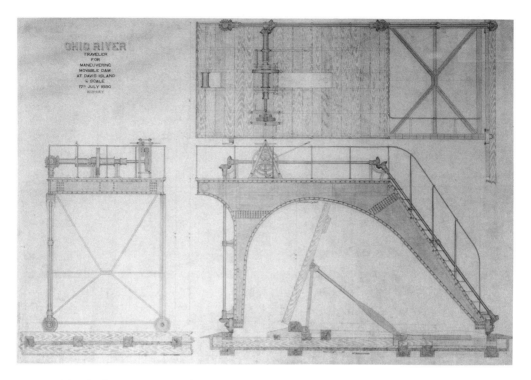

Traveler for maneuvering the Corps'
movable dam at Davis Island, patterned
after an 1852 invention by French engineer
Jacques Henri Maurice Chanoine.

Courtesy of Record Group 77,
Cartographic Division, National
Archives.

North American Review, reporting the popular view in 1821.[92] Six years later
a Mississippi congressman visited West Point to decry the "devastation and
misery" caused by the American "prejudice" against men of science.[93] Likewise when Ellet made plans to study in Europe, his father had a word of advice: "scientific knowledge" might handicap a young engineer, perhaps
leaving him "destitute of employment."[94]

Ellet, nevertheless, spent about a year in the advanced course at the Ecole
des ponts et chaussées, the first of three trips abroad. Returning to Philadelphia in 1832, he launched a unique career, tweaking the building professions
with unorthodox French ideas—bridges suspended from wires, reservoirs
on the upland tributaries of the Ohio and Mississippi that would regulate the
rivers and protect the heartland from floods. Ellet defiantly searched beyond
the British tradition, yet he remained, by training and choice, an empirical
builder outside the sphere of West Point. An apprenticed canal and railroad

surveyor, he was largely self-taught. His most famous work as a builder, the wire suspension bridge at Wheeling, Virginia, was a triumph of materials testing and construction know-how that drew little from theoretical science. His river reservoir scheme, although French inspired, ran aground on the army's plan to build a system of levees. Breaking with army science in 1853, Ellet published a book that challenged the War Department with bold, untested ideas about delta currents and tides.[95]

Ellet's American classmate at the Ecole des ponts et chaussées was closer to the theoretical bent of the Corps and West Point. Charles Storrow, bound for Paris at age twenty in 1829, heard the same kind of advice that Ellet had received from his father. Civil engineer Laommi Baldwin, Jr., told Storrow to strive for "practical knowledge," learning by observation.[96] Inventor Philip Taylor said an aspiring builder could safely dispense with science. Yet Storrow was drawn to hydrodynamics. A careful student of the French engineer-theoreticians—Bossut, Columb, Prony, Navier—he also studied Mahan. Historian Peter M. Molloy cites Mahan as an influence on Storrow's most lasting achievement, the Great Stone Dam at Lawrence on the Merrimack River, the largest hydropower dam of its time. Built in the late 1840s under the direction of Storrow and the former Corps engineer Charles H. Bigelow, the nine-hundred-foot dam was true to army hydraulics in several respects: it was stone rather than wood; it slanted and arched across the river so that high water flooding over the dam would crest in the middle of the channel; and the upstream face of the dam was slightly inclined because, as Mahan had explained, a perfectly vertical structure caused dangerous "whirls" in the river that eroded its bed.[97] Storrow also designed sluices, spillways, and a mile-long canal with enough power capacity to drive fourteen textile mills. Bigelow, a convert to lime burning, mixed *béton* for the rubble foundations.[98]

As late as 1842 it was still possible for West Pointers like Totten and Hughes to idolize "the most eminent French engineers" for their "science and profound investigation," but eighteenth century stereotypes no longer applied.[99] Ever-changing science-based engineering was becoming a worldly profession, a fusion of disparate techniques. Polytechnic savants now went to Britain to marvel at railroads and iron production. Hughes studied Dutch hydraulics. Alexander D. Bache, the West Point scientist who replaced Hassler at the U.S. Coast Survey, strived to combine British hydrography with French education, the German study of terrestrial magnetism, and Yankee innovations like the "American method" of fixing longitude by telegraph. Bache associates John D. Graham and Andrew A. Humphreys both won scientific acclaim through hydrographical surveys. Graham, a veteran boundary surveyor, was the first to discover and chart lunar tides on the Great Lakes.

Humphreys, a future chief engineer, challenged French hydrodynamics in his Mississippi Delta Survey. Meanwhile the private sector employed a "New England school" of hydropower experts—men like turbine inventors James B. Frances and Uriah A. Boyden, Massachusetts scientists and engineers who remained skeptically British in their dread of the mathematical laws that idealized hydrodynamics.[100]

By 1850 the divergent European traditions had merged in a kind of commercial project that showcased West Point construction and blended foreign technique—the federal lighthouse. Tall lights in hazardous settings were towering manifestations of America's commitment to commerce, a metaphor for national power. "A light-house," said *Putnam's Monthly* in 1856, "is an embodiment of national majesty, and, prouder than all, it is an untiring assertor of the common brotherhood and united humanities of the nations upon earth."[101] Stoic, picturesque, the lighthouse remains a romantic symbol, but its classic American beauty belies a cosmopolitan past.

The army began building lighthouses in an age of mounting dissatisfaction with the strict economy and low-bid contracting system of the U.S. Lighthouse Establishment, a Treasury Department bureau. In 1838 a board of navy inspectors made a list of concerns. While three towers in Maine were under construction in virtually the same location, there were dangerous stretches of coastline with no lights at all. At Bodie Island, North Carolina, a brick tower had shifted in sand, tilting the light. Elsewhere the builders mixed wood into mortar. Machinery rusted. Lamps were installed backward. Reflectors were dusty and bent. Of the 204 lighthouses and 28 lightboats in the federal system, the navy estimated that 40 percent of the lights had serious defects. "The whole lighthouse system needs revision," said nautical publishers Edmund and George W. Blunt, petitioning Congress for "strict supervision and an entirely different plan of operation."[102]

The first attempt at reform was to transfer some of the lighthouse work to the army. In 1831 the Corps began to receive a trickle of treasury funding for tower improvements, but Charles Gratiot, the chief engineer, preferred to build fortifications. By 1837 the War Office had spent only $10,792 of a $71,286 appropriation. Ten years later the Polk administration and Congress gave six unfinished projects to the topographical engineers. Finally, in 1852, Congress created an oversight Lighthouse Board. Three of the nine members—Totten, Kearney, and Lt. Edmund L. Hardcastle—worked for the War Department. Board member Alexander Bache brought top West Pointers into the program through a revitalized coast survey.[103]

West Pointers came in at a time when Europe, following Britain's lead, was discovering girders and I-beams, learning to build towers and bridges with iron. "One must go to England to appreciate the value of iron," said French industrialist Michael Chevalier, reporting from London in 1833. Iron frames were not only strong and fire-resistant, they could be prefabricated at a foundry and quickly assembled on site. Ready-to-assemble houses and wharves had already reached the British Caribbean. In 1838 the British inventor Alexander Mitchell adapted the technology to shallow water with a simple device: he fitted wrought-iron piles with broad-bladed screws. Now beach foundations could be twisted into the ground. At Maplin Sands Lighthouse near the mouth of the Thames, the sea swept through the open frame of Mitchell's first screw-pile tower. A second Mitchell lighthouse stood near Fleetwood, England, at a depth of twenty-eight feet. Americans took note when two West Pointers, topographers Hughes and William Swift, returned with drawings from England in 1841. Soon the topographical corps was improving Mitchell's design. Swift used wrought-iron stilts in the 1843 renovation of Black Rock Ledge Beacon, a skeleton tower southwest of Bridgeport. Lt. William H. Emory introduced the technology to the Franklin Institute with plans for an open-frame screw-pile pier that allowed the passage of current but blocked floating ice.[104]

Maj. Hartman Bache immediately saw potential for wrought-iron framing in wharves and lighthouse foundations battered by thick sheets of ice. A polished, artistic West Pointer from the first family of American science—a cousin of coast surveyor Alexander Bache, a great-grandson of Benjamin Franklin—Bache supervised lighthouse projects on three coastlines. His most outstanding achievement was one of the early American screw-piles, a stout platform at Brandywine Shoal in Delaware Bay. Completed in 1850, Brandywine Shoal Lighthouse stood forty-six feet above sea level, its keeper's house suspended just below the light. The project's most novel feature was an outer "ice-breaker" with thirty-seven piles bolted into a fence. Bache also experimented with lanterns and fog bells, and in 1859 he invented a remarkable wave-powered fog trumpet that could be heard for at least seven miles. Two energetic colleagues also used screw-pile construction. George G. Meade, the future Union commander at Gettysburg, developed the iron prototype for tall, pyramidal, hurricane-resistant towers along the Florida Keys. Fellow topographer Graham, Meade's brother-in-law, designed an early screw-pile pier and beacon for Chicago's North Pier. Lighted in 1859, it was one of first iron lighthouses with piped-in gas.[105]

Iron lighthouse towers previewed the skeletal framing of the American skyscraper, but before the age of steel the technology was uncertain. One cold night in 1856, for example, the lightkeepers at Cross Ledge, Delaware, awoke to find their wrought-iron platform uprooted by slow-moving ice.

Sand Key Lighthouse, designed by topographer George G. Meade and completed in 1853. Courtesy of the Historian's Office, U.S. Coast Guard.

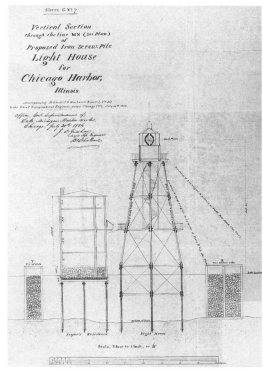

Plan of a screw-pile pier and lighthouse for Chicago, drawn by topographer John D. Graham, 1854.
Courtesy of Record Group 77, Cartographic Division, National Archives.

Four years later the West Pointer Charles Bigelow, formerly Storrow's assistant, took the blame for the faulty cast-iron pillars in Lawrence, Massachusetts, that brought down the Pemberton Mill. Yet the army's interest in iron continued to fuel innovation. Capt. Richard Delafield, a future chief engineer, designed the nation's first all-metal bridge at Brownsville, Pennsylvania. Meanwhile the West Point Foundry, a private firm headed by academy graduate Robert P. Parrott, turned out new kinds of rifles and cannon, and in 1843 the foundry casted and welded a hull for the first iron steamboat on the Great Lakes, the USS *Abert.* Later the Corps engineers proved the strength of rolled I-beams at the U.S. Assay Office on Wall Street (1853) and the wrought-iron Galveston Post Office and Courthouse (1861). Totten caught the attention of Europe with metal armor on fortifications. West Pointer William S. Smith, a civil engineer, helped design one of the world's first all-steel trusses, the railroad bridge at Glasgow, Missouri.[106]

Lighthouse construction kept pace with these innovations by mixing European ideas. Whereas the metal brawn of the towers was British inspired, the lamp itself, the scientific brain of an army lighthouse, was a French innovation. In 1822 a French physicist, Augustin Fresnel, invented a superior lantern with lenses and prisms that refracted a bright sheet of light. At first the West Pointers stood almost alone in support of these prism lenses. In 1830 the lighthouse office wrote a letter to a lamp maker in Paris, but a decade passed before Congress agreed to a test. At last in 1851 a military board of experts ruled that prism lenses were "vastly superior." The old washbasin-type reflecting lamp, the parabolic reflector, was "nearly obsolete."[107] Major Bache had already installed a large Fresnel lens and lantern at Brandywine Shoal. Meade invented a fuel-pumping system that improved the French design.[108]

French lights on British foundations were the sensational trial of West Point engineering, and no test was a steeper challenge than a flat rock south of Boston called Outer Minots Ledge. Hammered by swells and barely twenty-five feet in diameter at the lowest tide, it was, builders maintained, the most treacherous site in America, a ledge in a gauntlet of reefs exposed to the open sea. From 1832 to 1841 at least forty ships had wrecked on these rocks. Once a bark laden with hemp and iron had been shattered in half. When Congress authorized a beacon, an auditor in the lighthouse office, more worried about money than safety, balked at the expense of a traditional tower. A Mitchell-type screw-pile was the cheapest solution. A skeleton frame, moreover, would minimize the terrible shock of each crashing wave. In April 1847, the problem was laid before Capt. William Swift, the topographer who brought screw-piles to America at Black Rock Beacon. Known for

Brave lightkeepers board William Swift's Minots Ledge Lighthouse in rough seas, 1850. Battered by swells, the structure collapsed in 1851. Courtesy of the Prints Department, Boston Public Library.

his map of the "great desert," Swift had been to the Rockies with Long, worked on the Chesapeake and Ohio Canal, and mapped for the coast survey. Minots Ledge was his last army project. A grand experiment in iron construction, it left him a broken man.[109]

Swift's lighthouse, completed in 1850, resembled a giant spider on nine iron legs. Screwed and cemented into the rock, it stood seventy-five feet on ten-inch-diameter piles. A five-ton capping platform held a turretlike lantern room with a new Fresnel light. Twice during construction the unpredictable swells had tossed men and machines into the ocean. Miraculously, no lives were lost.

Even while Minots Ledge was being hailed as a landmark of iron construction, the men who attended the light lost faith in their vibrating roost. Waves crashing over the deck pitched the frame like a boat and pounded it like a drum. After nine terrible months the chief lightkeeper resigned. His replacement soon sent a letter of protest, but Swift, publishing a lengthy defense of the structure, insisted that the iron was sound. Then, in March 1851, a furious gale smashed a lifeboat and twisted some of the brackets. Five weeks later, on Wednesday, April 16, high winds and sixty-foot swells extinguished the light. A newspaper reported "the highest tide ever known in Boston."[110] As surf rolled into the city, flooding streets, thousands ran to the dockyards to gasp at the boiling ocean. That night around eleven o'clock a monstrous wall of water smashed through the iron lighthouse, cracking its central support. One by one the eight remaining piles slowly bent backward and snapped. Bravely, the two lightkeepers corked a message into a bottle, strapped on life jackets, and lowered themselves to the surf. One fought his way to some rocks, dying of exposure and shock. His dead colleague eventually washed up on the beach, nearby.

Soon after the storm the hapless promoter of screw-pile construction rowed out to the horrible site. Swift, now semiretired, made a sketch of the wreckage—the tower on its side in the surf, the rock with nine iron stumps. It was a cruel day in Massachusetts, a bitter end to a tarnished career.

Immediately the army made plans for a second lighthouse—a conical masonry structure, a veritable fort. The obvious choice was fort engineer Joseph Totten. Combative and powerfully built, he fit the part of the soldier. As chief engineer and a senior member of the U.S. Lighthouse Board, he could be shrewdly political, yet, in devotion to the methods of science, his integrity was beyond reproach. Totten insisted on a massive tower on the same deadly rock. He pored over plans of the great British beacons, gathering information on the Eddystone Lighthouse and others, but soon Totten concluded that Minots Ledge with its small exposed surface was in a dangerous class by itself. The solution was to combine the mass of cut stone with the tensile strength of wrought iron. Seven large foundation blocks would keep the weight of the tower close to the rock. The solid base would narrow into a hollow cone with a staircase, a house, and the light. To hold the structure together and anchor it to the rock, the plans called for two-inch galvanized bolts and an internal frame of connecting rods in the same position as Swift's screw-pile stilts. Totten chose Capt. Barton S. Alexander, one of his own engineers, to supervise construction. A patient builder, Alexander was smart enough to know when to rethink Totten's plan.[111]

Construction began in June 1855. Working from plans and two scale models, the builders removed the old pilings, redrilled the holes, and shaved the rock to a level surface two feet below lowest tide. While Alexander erected a tall iron derrick, shiploads of Quincy granite—about thirty-five hundred tons—were delivered to a nearby island and then cut into blocks. The sea, as before, was a brutal companion. On January 19, 1857, a ship loaded with cotton crashed into the site, chipping the ledge and destroying the derrick. Cracks in the rock forced Alexander to redesign the foundation stonework.

Starting over that spring, crews ringed the ledge with hundreds of sandbags, mopped it dry with giant sponges, and began setting the granite into Portland cement. Three years later the last stone was in place. Capped with a domelike cupola and a fixed Fresnel lens, the completed lighthouse, twenty-three hundred tons of granite, was the Corps' most massive tower. It rose from a 30-foot base to a height of 114 feet. At sunset on November 15, 1860, as bonfires marked the shoreline and sky-rockets shrieked overhead, the second Minots Ledge beacon sent out its first flash of light. Now operated by the U.S. Coast Guard, the granite tower, together with Smeaton's Eddystone Lighthouse, has been called "one of the two foremost wave-swept lighthouses in the world."[112]

A failure, a success, a beacon of technology transfer both British and French—Minots Ledge Lighthouse was much like the Corps itself. France taught the organization to build grandly at public expense, to test iron and concrete, to promote standardization, and to make a meticulous science of currents and tides. The rediscovery of British construction brought in another tradition. As Britain eclipsed France as a center of innovation, the army, especially the topographical bureau, reached out to gifted mechanics who modernized iron production and built spectacular towers.

Did West Point's cosmopolitan science radically change civilian design? Probably not. Although engineer-officers virtually monopolized the largest, most ambitious federal construction projects, the army was forced to rely on the same kind of building mechanics that served the nation at large. Many of these field-trained builders and craftsmen resisted West Point innovation. Even in the firearms industry a fear of militarism crippled the West Point attempt to modernize the factory system. As Merritt Roe Smith explained in his study of the national armory at Harpers Ferry, the civilian work force found army discipline and regimentation repulsive and French ideas about efficiency through standardization were widely denounced as insults to the prideful independence of America's laboring class.[113]

Another limitation of the scientific approach was the ambiguity of the word "science" itself. Science, as many savants used the word, was a probe of the mystery of nature, a nonmaterialistic pursuit. But engineers were preoccupied with practical applications. As historian Edwin T. Layton, Jr., explained, most engineers shunned "idealizations" that allowed the physicist or true theoretician to describe something in nature that was hard to measure or test.[114] Wary of grand abstractions, fascinated by natural laws but finding them hard to apply, builders and inventors made do with the methods of science: the laboratory setting, the scale models, the precise instruments, the scientist's mode of communication through reports and technical journals.

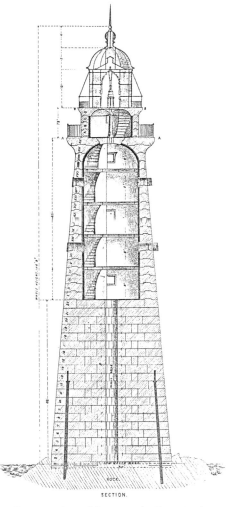

SECTION.

Cross-section of the second Minots Ledge
Lighthouse, 1860. Courtesy of Record
Group 77, Cartographic Division,
National Archives.

Second Minots Ledge Lighthouse.
Designed by Joseph G. Totten and
completed by Barton S. Alexander in
1860, the 114-foot beacon remains one
of the world's most spectacular towers.
Courtesy of the Historian's Office,
U.S. Coast Guard.

Still, the West Pointers had towering expectations for science. An important buzzword in the nineteenth-century army, science embodied disparate objectives and values: the love of order, the promise of technological progress through the conquest of nature, the romance of warfare as a Napoleonic chess game, a link to the world of Vauban. Science, in the language of army construction, was also rational planning. A yardstick of fairness in government, science measured the national interest against the rights of the states. Ultimately, that utilitarian application of science did as much for engineering politics as it did for engineering design.

4. Engineers Diffuse Innovation

In America few lasting ideas have sparked from the mind of the genius inventor alone in the shop or the lab. Our technology has slowly evolved, much of it inspired by Europe. Americans, nevertheless, take pride in patriotic conceits about self-reliance through innovation. Henry Steele Commanger and many others have written eloquently about an American culture of action and instinct that won independence from Europe through new kinds of tools and machines. In 1978, Daniel J. Boorstin agreed that American advances in manufacturing and transportation were a technological revolution that homogenized our society, helping us repudiate things European. Pistols, printing presses, public highways and railroads, farm and household implements stamped out by an "American System" of factory production through interchangeazble parts—these and other innovations, Boorstin maintained, were a source of national identity, a bonding force that broke down regionalism and the barriers of social class.[115]

Ironically the so-called American System of factory production was a military concept from Europe. So was the central idea behind the Whig Party's American System of public aid to the railroads. Both systems were ideas more French than British, and both were championed in America by schooled technologists who had little in common with the rustic jack-of-all-trades.

Far from homogenizing the culture, military innovation, coming as it did from perceived outsiders and despots, antagonized an American work force rooted in craft tradition. Perhaps that antagonism explains why so little has been written about the military-scientific connection that helped the nation mature as an industrial power.

Neglect of that West Point tradition has distorted our historical thinking in subtle but significant ways. Although scholars like Monte Calvert and Joel Tarr have written with great insight about engineering "cultures" in conflict, the literature on early construction downplays factionalism. Stressing unity, the history books mostly rely on what sociologists have called a "functionalist model" of the engineering profession as a tight-knit community of like-minded women and men with common values and goals. There is little room in the functionalist model for builders who saw themselves as a scientific alternative to the craft tradition. Instead of defining engineering by "function," it might be more useful to understand professionalization as the "process" through which builders from different backgrounds vied for status and recognition while competing for big assignments and adding to the cosmopolitan ferment of American styles and techniques.[116]

If professionalization was a process through which scholars, soldiers, surveyors, and craftsmen all interacted as builders, it stands to reason that different professional cultures had different sources of information. Books, for example, were revered by the builders with access to the best libraries in Europe. But many historians concentrate on the word-of-mouth tradition emanating from Great Britain. Slighting the value of the plans and reports so cherished by bookish West Pointers, the literature devalues the scientific mode of expression that helped the Corps establish its niche. "The printed or written page speaks only to those who know already the kind of thing to expect from it," wrote Carlo M. Cipola in a history of technology transfer. "Even today blueprints are considered inadequate to transmit full information, and when a firm buys new and elaborate machines it sends some of its workers to acquire, directly from the builders, the knowledge of how to operate the new machine."[117]

While it is certainly true that person-to-person contacts were essential to technology transfer, the manuscripts exchanged by soldiers—books, treatises, plans, maps, journals, and scientific correspondence—recorded and preserved technological innovation, passing a kind of information that was too precise for the spoken word. Careful West Pointers refused to proceed without a stack of plans and reports. At the Potomac Aqueduct, for example, where the army was studying ways to bring the Chesapeake and Ohio Canal from Georgetown to Alexandria, Virginia, Maj. William Turnbull, a topographer, searched in vain for drawings or a good descriptive memoir of London Bridge. Finding no British publications, Turnbull based his aqueduct piers on two French designs. Likewise at Brandywine Shoal the engineers worked from a British patent, Swift's lighthouse reports, and imported plans of screw-pile foundations. West Point encouraged the paper exchange by creating a department of drawing, and in 1837, under the able direction of the Italian-trained landscape artist Robert W. Weir, the academy became perhaps the first American college to offer an advanced course in "tracing," "copying," "topographical delineation of rocks and hills," "the art of shading geometrical figures," and "aerial perspective." [118] Mahan, an excellent drafts-man, returned from France with diagrams for lessons on architecture and industrial design. [119]

Another source of the West Point paper exchange was respect for applied mathematics. Math, a method of inquiry, was also a system of logic. In 1834 the *American Quarterly Review* referred to the French advance in algebra and descriptive geometry that was replacing the deductive "Newtonian method of investigation" at the West Point academy. [120] Whether engineers built fortifications or not, Totten insisted that "extended mathematical attainments" were the mark of a qualified builder. [121] Major Long proved the point by using algebra and trigonometry to calculate the strength of a timber truss. The technique, at first rejected by railroad builders, led to the patented timber-frame "Long truss," a popular bridge design. [122] Long also helped Capt. Andrew A. Humphreys launch his statistical study of the Mississippi Delta. After a comprehensive investigation of French, German, and Italian hydraulics, Humphreys, ably assisted by Lt. Henry L. Abbot, developed a "universal formula" to measure the carrying capacity and sediment load of delta-forming rivers. The Humphreys-Abbot equation was then tested, debated, and ultimately rejected in Europe. Although Long had never visited Europe and Humphreys had been there just once, mathematical innovation had crossed and recrossed the Atlantic through the international language of the engineering report. [123]

Thus the statistical, mathematical, hydrographical report was in itself a kind of technological innovation that brought West Pointers into construction without straining the national preference for empirically trained engineers. While historians make a point by stressing the limits of the paper exchange, it is too much to say, as one writer maintained, that "men building things [before 1860] proceeded in ignorance of how much a locomotive could pull, how much a beam could bear, how much water flowed through a channel at a given point." [124] West Pointers knew exactly these things. Like the polytechnic savants, like Thayer and the Frenchman Crozet, army builders reversed the traditional order of American technical training, stepping from theory to practice, from the principles set down in writing to the challenge of design in the field.

The value of that academic tradition was real but hard to assess. No book taught engineers how to manage a complex assignment or straddle the growing debate over who should fund public works. That education began in 1824 when the Corps and its topographical bureau, now servants of Congress, moved apolitical planning into the political world where builders defended their plans.

Objects of National Pride

David Stevenson said Yankee construction was ingenious but remarkably coarse. Reporting on America in 1837, Stevenson, a visiting Scotsman, tried to explain why a nation so good at commerce would ignore its principal ports. At New York, for example, the wharves were exposed to the ocean without even a cheap coat of paint. Shanty dockyards were "temporary" and "far from agreeable," and the Scotsman was sure that "the smallest of the post-office packet stations in the Irish Sea has required a much larger expenditure of capital, than the Americans have invested in the formation of harbour accommodations for trading vessels along a line of coast of no less than 4000 miles, extending from the Gulf of the St. Lawrence to the Mississippi." [1]

Stevenson was wrong about the expense of public dockyards and harbors, but his 1837 report, published in London the following year, supported an egalitarian myth: the United States, a frenetic republic, was a nation too restless for science and too jealous of federal power to fund large public works.

Today a vast literature on construction in Andrew Jackson's America builds from that same idea. An era of state promotion and private construction, the Age of Jackson, most scholars agree, was a time when a Congress beaten back by the veto did little for water commerce. George R. Taylor's seminal work on *The Transportation Revolution* (1951) minimizes aid to rivers and harbors, calling it "indirect" and "much smaller" than the federal investment in turnpikes. [2] Edward L. Pross, an authority on waterway bills, agrees that federal funding was "far too small an amount for important results." [3] Lawrence M. Friedman's *History of American Law* (1973) and Jamil Zainaldin's *Law in Antebellum Society* (1983) reinforce that interpretation by neglecting the river debate and omitting the hard-fought compromise that was, perhaps, the most critical piece of nineteenth-century waterway legislation, the General Survey Act of 1824. Likewise many economic historians have misread Andrew Jackson's attack on big construction programs. Some say Jackson killed an era of national projects with a veto of federal aid to a Kentucky highway, the Maysville Road. After 1830, Carter Goodrich explains, "Congress appropriated no money for new improvements until after the Civil War."[4]

Congress, in fact, spent at least $43 million on waterway projects before 1861—the cost of six Erie Canals.[5] That expenditure peaked during Jackson's eight years in the White House and again, briefly, in 1852. Although small by later standards, the first era of national projects created a vast civil-works jurisdiction for the Corps and the topographical bureau. It funded a half century of intense hydrographical investigation. It fused science with transport planning and allowed about seventy top West Pointers to manage water resources: to survey every major river, identify hazards, propose projects, supervise construction, and guide a Congress too weak and confused to design an intelligent plan.

Certainly the $43 million was a far cry from Stevenson's estimate. Had the Scotsman landed at Philadelphia instead of New York, entering the Delaware River instead of Long Island Sound, he would have seen tax dollars at work. At Cape Henlopen, Delaware, he would have passed a half-finished breakwater and ice pier, the army's man-made harbor of refuge, one of the largest in the world. Upstream was a chain of rectangular piers that were among the first river projects to receive a direct appropriation from Congress. Some of these timber structures predated the Revolution. When the new government had assumed the right to levy a tax on shipping, Congress, the shippers insisted, had inherited the obligation to maintain safe navigation. In 1802 the House provided $30,000 for an ice harbor at Chester, Pennsylvania, half the amount requested. Massachusetts and Maryland made a similar arrangement to siphon some of the tariff for harbor and channel improvements. The port of Savannah used Treasury funds to buy a dredge boat, remove a wreck, and seal off the Back River.[6]

The legality of this kind of assistance remained a gray area between two readings of the Constitution. That document said nothing about navigational works per se, but the Federalists saw a solution. Returning to Hamilton's brilliant campaign on behalf of a national bank, the Federalists said government had "implied powers," a broad authority to create financial institutions, taxing authorities, a lighthouse board, and other agencies that seemed vital to the general welfare. Congress could remove a hazardous reef just as surely as the president as commander in chief could outfit a campaign against pirates. Both protected public safety. Congress, moreover, could "regulate commerce among the several states."[7] If "commerce" included navigation—and Hamilton insisted it did—then here was an invitation to build or at least subsidize all kinds of water projects. Finally, the Federalists fixed on the constitutional provisions for a peacetime army. These provisions *implied* the authority to map strategic waterways, dredge harbors for the navy, move troops through canals, train officers in a national academy, and use these army experts to direct public works.[8]

Democratic Republicans read a different Constitution. Jefferson's people, they drew strength from the South and the heartland, and their leaders came to believe that almost any spending proposal threatened the farming republic. Even science, engineering, and internal improvements—the things Jefferson loved—might feed the dens of corruption where Federalists were grasping for power. When, in 1796, a House resolution endorsed a postal road from Maine to Georgia, Jefferson objected. "I view it as a source of boundless patronage," Jefferson wrote Madison. Road building would quickly become "a bottomless abyss of public money," a cash cow for the friends of Congress.[9]

Trading from timber dockyards, Philadelphia, 1858. Courtesy of the Library Company of Philadelphia.

Coal depot at Richmond on the Delaware
River, about 1830. Timber piers on the
Delaware River were among the first river
construction projects to tap federal aid.
Courtesy of the Historical Society of
Pennsylvania.

A close look at Jefferson's protest shows that his position on the postal
road was less an ideological statement than a political attack on the monied
gentry of New York and New England, the Federalist opposition. But Jefferson was not above an indictment of Congress to make an ideological point.
Distant authority, bloated government, sinister commerce—these Jeffersonian themes drove a wedge through internal improvements. Breaking the
Lockean dream of government by democratic consensus, they incited a fear
of corruption that toppled President Adams but haunted the new party in
power. In 1801 Jefferson faced a choice between a strict reading of the Constitution and the pro-expansionist policies that moved Americans west. How
could government unite the Republic and open agricultural markets while
preaching fiscal restraint? How could planners and builders bring order to
internal improvements without tipping the balance of power that preserved
the rights of the states?

The safest course seemed the most indirect. Jefferson permitted small harbor and lighthouse improvements in the name of public safety. In 1802 he
objected to money for piers in the Delaware River but refused to veto the
measure. Citing the need to deliver the mail, he allowed the proceeds from
public land sales to fund a highway to Ohio, the future Cumberland Road.
Jefferson also pushed for a road to New Orleans and, turning to maritime
commerce, he endorsed Ferdinand Hassler's plan for a U.S. Coast Survey.
All the while the President argued the need for a constitutional amendment

so that Congress, "in time of peace," might invest in "roads, canals, rivers, education, and other great foundations of prosperity and union," not piecemeal subsidies for private builders but a fair and comprehensive "system of improvement."[10]

The famous *Report of the Secretary of the Treasury on the Subject of Public Roads and Canals,* the 1808 Gallatin Report, spelled out what Jefferson might have meant by a "system" of internal improvements, and it showed a new tolerance among leading Republicans for implied constitutional powers. Treasury Secretary Albert Gallatin of western Pennsylvania was a vigorous supporter of the Cumberland Road and a leader of the frontier campaign for access to the coastline. His report pieced together dozens of ambitious proposals—frontier highways, a chain of canals along the seaboard, a canal across New York from the Hudson to Lake Ontario, and shorter canals around Niagara Falls and the falls of the Ohio. Boldly, it called for "direct aid from government," an investment of $20 million in $2 million installments over the next ten years. The plan also reflected some of Jefferson's main concerns. Where Hamilton and his friends had emphasized Atlantic seaboard commerce, Gallatin, looking west, wanted roads across Appalachia and beyond to Louisiana. Gallatin, moreover, stressed the virtues of farming. A good network of roads and canals would "enable every industrious citizen to become a freeholder, to secure indisputable titles to the purchasers, to obtain a natural revenue, and above all to suppress monopoly."[11] While admitting, for Jefferson's sake, the possible need for a constitutional amendment, Gallatin also suggested a more immediate action: government surveyors

Joseph G. Swift, chief engineer, 1812–1818.
Courtesy of the Office of History, U.S.
Army Corps of Engineers.

could work on an atlas of feasible projects, a grand blueprint for federal aid.[12]

Congress tabled the costly proposal, but Gallatin's ambition lived on. From 1808 to the present, while the debate over rivers and harbors has often seemed hopelessly deadlocked, a faith in the wisdom of science has permitted the experts to plan. Gallatin was sure that educated men like Latrobe and Fulton could rise above the petty concerns that derailed magnificent projects. Almost two centuries later, looking back on what was actually built, the blueprint of 1808 seems both wise and naive—wise because federal bureaus did push for waterway planning; naive because planning, never impartial, was torn by the rival traditions that divided American builders and precluded a lasting consensus on federal public works.

1. A Good System of Roads and Canals

The Corps, try as it might, could never escape the fact that army engineering was often the political science that forced builders into complex disputes. Engineers considered themselves tough-minded problem-solvers, men who cut through conjecture by staying true to the facts.[13] Congress, however, increasingly relied on the builders to speculate, negotiate, bridge political conflicts, and use engineering in ways never discussed at West Point.

One engineer who learned to make political choices was the first graduate of West Point and second chief of the Corps, Gen. Joseph G. Swift. At first Swift was determined to keep his distance from politics. He was so determined, in fact, that in 1810 Swift took his troops to an island off North Carolina to sit out a state election. Politics, Swift observed, was a taboo subject in an army mess hall, yet his family and most of his friends had supported President Adams, and his own fast-track career exploited the inner circles where statesmen talked public affairs. Once at a Washington party Jefferson asked him bluntly: "To which of the political creeds do you adhere?" Swift was forced to admit his link to the Federalist party. "There are many men of high talent and integrity in that party, but," the president said, "it was not the rising power." [14] Taking the hint, Swift was soon closely allied with Madison and rising Republican stars like Henry Clay and John Calhoun. When Calhoun ran for president in 1824, Swift supported the campaign with a pamphlet—a costly mistake. Calhoun lost the election. The friends of President-elect John Quincy Adams saw to it that Swift was ruined in New York. Isolated and nearly bankrupt, the engineer moved to Tennessee and fell in with Andrew Jackson. His subsequent career was a series of high appointments—Great Lakes harbor superintendent for Presidents Jackson and Van Buren, Canadian envoy for Harrison and Tyler, American Whig Party delegate in the early 1850s, and in 1856, an antislavery spokesman for the new Republican party.[15]

From Adams Federalist to Jackson Democrat to Henry Clay Whig, Swift's career spanned a turbulent era of realignment, a time when the political loyalties of the constitutional period broke down and reappeared in the 1830s as a struggle between states-rights Democrats and pro-Union Whigs, the second two-party system. Swift stayed afloat by shifting with the nation at large. An ambivalent politician, the one constant in his public life was passionate support for a Gallatin-like system of federally funded improvements. He even began a book on the subject. Like engineer Robert E. Lee, a thinking Democrat who, his biographer said, "argued downrightly for internal improvements," Swift and his fellow West Pointers chiefly wanted to build.[16] Party ideology was never as strong as the sense of profession that bound engineering to Gallatin's cause.

The power of Gallatin's vision helped focus Swift and the Corps during a long span of relative peace between the War of 1812 and the Mexican War. It began with an exuberant time of patriotism and public improvement that a Boston newspaper called an "era of good feelings."[17] If one man captured the mood, it was Henry Clay of Kentucky. The tall, high-spirited planter and entrepreneur had, as Speaker of the House in 1815, returned the attention of Congress to rivers and navigation by challenging the British right to trade freely on the great Mississippi. Three ideas cemented Clay's coalition. First was a high tax on imports, a tariff, to protect American products; second, a network of internal improvements; third, a strong federal bank. Clay sold the Hamiltonian program as an engine of centralization. Aiding business while protecting the laboring classes, it would give "power and strength to our Union by [establishing] new ties of interests, blending and connecting together all of its parts, and creating an interest with each in the prosperity of the whole."[18] Clay called it "The American System."

An early test for Clay and the nationalist program was Calhoun's elaborate plan to shift proceeds, or "bonuses," from the Bank of the United States into a fund for internal improvements. Known as the Bonus Bill, it moved slowly through Congress in February 1817. Congressman Thomas Wilson of Pennsylvania hoped the construction fund would "perpetuate the republic" by fostering "an indissoluble community of interests, habits, and attachments."[19] For Congressman Henry St. George Tucker the issue was defense. Recalling "the embarrassments of the nation during war," he blamed a weak navy, an exposed coastline, and especially "the want of an internal water communication."[20] Tucker, a Virginia cavalry man, said the constitutional authority to declare war and raise an army gave Congress the power to finance public works.[21]

Military aspects of the Bonus Bill implied an expanded role for Jefferson's fighting savants. Calhoun, praising the Corps, called for "a good military system of roads and canals," a system surveyed by West Pointers and directed at the federal level by a transport planning board.[22] It was hardly a new idea. New Yorkers had already created a blue-ribbon state commission to promote the Erie Canal, and Virginia, with Latrobe's help, was searching for a qualified builder to coordinate public works. Another model was Brig. Gen. Simon Bernard's Board of Engineers for Fortifications. Created in 1816, the board studied the coastline, reporting on navigation as well as defense. The ten topographical officers also used transport planning as they mapped ways to move troops. In 1817, for example, Maj. Long saw that a system of short canals and minor channel improvements could connect Lake Michigan to the Mississippi River. That year Major James Kearney fed the southern campaign for internal improvements with a plan for a deep-water canal from Chesapeake Bay to Albemarle Sound.[23]

Engineers argued the link between waterways and defense, but the campaign for federal projects still mired on legal ground. President Madison vetoed the Bonus Bill on March 3, 1817, his last act in office.[24] A constitutional purist, Madison said the power to regulate commerce did not give Congress a license to fund overtly commercial construction. There was probably more to this famous veto than the president was prepared to admit. The bill, its critics maintained, was the kind of spending program that Americans feared the most—a classic pork-barrel package, an attempt to dole out the federal treasury without a central plan. Calhoun had agreed to reward each state in proportion to its population, and he further weakened the measure with a clause that made Congress beg for consent from the states. In the end even Gallatin condemned the construction fund. Still, the death of the Bonus Bill was a shock to the Clay coalition and a personal defeat for Calhoun. Pragmatists now realized that there would be no progress on internal improvements until Congress created a board of transport planners to share power with the executive branch.[25]

The failure of national planning crippled federal action for the next seven years. From 1817 to 1824, the Monroe administration readily admitted the need for public financing, yet it insisted on state control. Congress resorted to indirect Jeffersonian tactics. It extended the lighthouse and pier program. It allowed the army to work on the Cumberland Road. It funded waterway exploration and asked Gallatin to compile statistics on French and British canals. In 1820 the Corps received $5,000 to study steamboat navigation on the Ohio and Mississippi. Again in 1822, when the army confirmed the need for a Delaware breakwater, Congress spent $22,000 on a feasibility study. Constitutional limitations forced the private sector to seed public construction through state and municipal ventures. New Yorkers, twice denied federal assistance, raised $7 million in state revenues and bonds for the 363-mile waterway from the Hudson to Lake Erie.[26]

Suddenly, as the Erie Canal neared completion in 1824, the waterway issue caught fire. Philadelphians made a bid for the western trade by proposing a link to the New York canals from Lake Seneca to the Susquehanna River. Anything seemed possible, even a passage to Asia. "Seventy-five miles of canals is all that is needed to give us full water communication with the Pacific Ocean," said one pamphleteer. In 1824 construction began on a locked canal across the Delaware peninsula. Virginians pushed a larger enterprise, the Chesapeake and Ohio. New Jersey, Ohio, Indiana, and Illinois launched ambitious projects, and Pennsylvania hoped to command "the whole inland trade."[27] What was needed, according to a bill taking shape in the House, were "surveys, plans, and estimates, to be made of the routes of such roads and Canals as he [the president] may deem of national importance in a commercial or military point of view."[28] Clay went further. Attacking Monroe and the states' rights position in a long speech on internal improvements, the Kentuckian, now a presidential candidate, insisted that Congress not only had the right but, indeed, "a great national duty" to open the veins of commerce that bound the East to the West.[29]

The U.S. Supreme Court was quick to respond. On March 2, 1824, about seven weeks after Clay's windy speech, Chief Justice John Marshall made a popular decision in a pivotal case, *Gibbons* v. *Ogden*. The case stretched back to 1808 when the New York legislature had granted steamboat monopoly rights to the inventor Robert Fulton and his backer Robert R. Livingston. In 1811 the Fulton-Livingston company had extended its control to the lower Mississippi. A portion of the New York business was leased to Aaron Ogden who, claiming an "exclusive privilege" to run steamers from New Jersey to Manhattan, secured a restraining injunction against his tough competitor, Thomas Gibbons.[30] Gibbons sued and lost in New York. Undaunted, the ferryman appealed with the help of two famous attorneys—William Wirt, the U.S. attorney general, and Daniel Webster of Boston, now serving his third term in Congress. Soon all eyes were on Webster. A fiery orator in his prime, Webster denounced the steamboat monopoly as a threat to the spirit of *E pluribus unum,* the belief that the separate states could work efficiently as a unit, the ideal of the federal state.[31]

Webster reduced *Gibbons* v. *Ogden* to a straightforward question of national supremacy. No state monopoly, Webster maintained, could infringe on the authority of Congress to regulate trade. The Marshall court agreed. "The power over commerce, including navigation, was one of the primary objects for which the people of America adopted their government," the chief justice explained. That power barred states from regulating or otherwise interfering with "the deep streams which penetrate our country in every direction" and "pass through the interior of almost every State in the Union." [32] Did the power belong *exclusively* to Congress? Could the states, for example, pass pilotage laws? Could they regulate some kinds of fares and tax some kinds of products? These points remained open debates in 1824. Still, the immediate impact of the case was a blow for laissez-faire. Crushing state monopolies, it freed investment and launched an era of fierce competition among hundreds of steamboat lines. *Gibbons* v. *Ogden* also sent a message to Congress: the great thoroughfares of inland navigation were federal jurisdiction. [33]

The steamboat monopoly case helped Congress avoid constitutional restrictions. In February 1824, while Webster was addressing the Court in the dimly lit basement of the U.S. Capitol Building, the House was in session upstairs. Congressmen Joseph Hemphill won Monroe's tacit support for a road and canal planning bill, a call for a presidential commission with surveyors attached to the Corps. Hemphill, a Philadelphia Federalist, had risen to power in the House as a vigorous promoter of turnpikes. He considered the planning proposal "the precursor to all future improvements," an appeal to "scientific men" and a search for "accurate knowledge." Calhoun put the matter more bluntly. "It became necessary," he claimed, "to advert our political system." [34] After a thirty-five-year stalemate between state and federal interests, the Corps, said Calhoun, would help the nation decide. Still, there was stiff resistance from the cities and states that already had boards of surveyors to plan roads and canals. Congressman Philip P. Barbour of Virginia, a Jeffersonian, said "matters of internal regulation" were "municipal powers," and he predicted that any "unequal" or "disproportionate application of moneys" would surely cause discontent. [35] On the other end of the political spectrum were those who believed the Hemphill proposal did not go far enough. Sen. Thomas Hart Benton of Missouri, a loud voice for internal improvements, insisted that Congress had the power to build without help from the executive branch. [36]

In the end the planning proposal carried with support from a unified West. Together in the House and Senate there were sixty-five votes from the nine states west of Pittsburgh. Sixty-three endorsed Hemphill's plan. New

York and Virginia, the two states already well along on their own public works, generally rejected the measure, but strong backing from Pennsylvania easily tipped the balance. The final vote in the House was 115 to 86. The Senate passed the bill 24 to 18.[37] On April 30, with Monroe's signature, the bill became "An Act to procure the necessary Surveys, Plans, and Estimates, upon the subject of Roads and Canals"—the General Survey Act of 1824.

The General Survey Act was a far-reaching compromise that took advantage of the commerce clause but compelled Congress and the executive branch to share political power. Shrewdly, the act stressed projects "of national importance," and it invited the chief executive to say which projects were more worthy than others. The president would employ "two or more skilful engineers, and such officers of the corps of engineers . . . as he may think proper."[38] In May 1824, Monroe appointed three distinguished builders to a Board of Engineers for Internal Improvement. With an initial appropriation of $30,000, the board, working through the army, began surveying an ambitious canal dear to the friends of Monroe—the Chesapeake and Ohio. Congress then extended the waterway program with two appropriations: $75,000 to clear sand and debris from the Ohio and Mississippi; $40,000 for pier improvements at two sites. These were the first army-directed water-construction acts in federal law.[39]

The events of 1824 were a middle path between rival positions. While Hamiltonians held out for direct aid to corporations and Jeffersonians continued to fear a "pernicious copartnership" between financiers and the executive branch, Congress, setting aside these issues, agreed to share power with experts.[40] A board of engineers would study canals and a highway system. A topographical bureau would find ways to improve dangerous rivers. Lighthouse boards and a coast survey would resume the harbor investigations that aided the maritime trade. But what if the experts wanted too much for the nation? How would government contain the ambition of its own magnificent plans?

2. "Great, Magnificent Government"

John Quincy Adams cheered the nationalist program. A convert to federal planning, the sixth president made "the spirit of [internal] improvement" a theme of his first annual message. Few men had more faith in government science. Adams wanted a national university in Washington, D.C., a national observatory, a professional naval academy, a new patent system, and a voyage

Simon Bernard, former aide-de-camp to Emperor Napoléon I. Banished from public life after the Battle of Waterloo, Bernard sailed to America in 1816. Courtesy of the Casemate Museum, Fort Monroe.

of discovery to the mouth of the Columbia. He also called for expanded civil-works training for the "meritorious" engineers at West Point.[41] Where Gallatin, Clay, and Calhoun had asked merely for roads and canals, Adams, a Harvard man with strong ties to the French enlightenment, envisioned a "great, magnificent government"; and in 1825 he proposed an executive department of science to educate the Republic and direct public works.[42]

In retrospect we can see that most of what Adams proposed was politically doomed from the start. Americans, still wanting practical knowledge, were not about to build a federal observatory or send the rich to college at public expense. No science agency won even a nod from Congress, but in its place an energetic Frenchman and a few assistants, the U.S. Board of Engineers for Internal Improvement, worked for the same ideal.

The Corps dominated the new board for internal improvement. Two of its three original members were Corps engineers, and the third engineer, civilian John L. Sullivan, resigned after a year. The ranking officer was fort engineer Simon Bernard, one of Napoléon's protégés. Board member Totten escorted Bernard, translated for him on occasion, and tirelessly defended his projects. Bernard and Totten began by reviewing rival plans for a Louisiana highway, and turning to navigation, they studied a link from New Orleans to Lake Pontchartrain. Quickly the surveying of waterways became

Joseph G. Totten, the future chief
engineer. Courtesy of the Office of
History, U.S. Army Corps of Engineers.

the board's primary focus. During seven years of joint command, 1824 to
1831, Bernard and Totten sent about thirty-five officers and at least eight
civilians on more than eighty waterway assignments. In 1826 they helped
Congress piece together the first omnibus river and harbor act—$86,000 for
improvements at twenty sites. Bernard and Totten also studied at least
twenty-eight proposed canals, postponing or rejecting twenty. By 1829 the
board's top four canals had received nearly two million federal dollars.[43]

Canal feasibility studies were an early test of the Corps' ability to influence
public policy through ideology and science. French and British waterways
became models for public projects that exceeded private resources. Bernard
and Totten spoke of a fiscal responsibility higher than "the standard of
money," making a distinction between canals that depended on public assistance and those that did not.[44] Some improvements were "like all commercial speculations." But elsewhere "the revenue from a canal may be much
less than that on ordinary investments, and yet the benefits amount to much
more." Canals were objects of national pride. "Revenue," the engineers
maintained, was "a secondary object."[45]

A case in point was the fourteen-mile cut that became one of the world's
busiest waterways, the Chesapeake and Delaware Canal. Bernard and Totten, working with two civil engineers in 1824, had rejected a popular plan
and recommended in its place a shorter but more costly project. The army's
construction estimate, $1.3 million, sent investors into shock. The canal com-
pany panicked. Its stock plunged. But engineers called the project essential,

and with $300,000 from Congress in 1825, construction proceeded. Four years later the completed canal was critically over budget. Again, the army gathered technical information that helped the stockholders lobby for funds. When the improvement board sent Major Long to examine the hazardous western approach to the canal, shallow Back Creek, canal president Robert M. Lewis "politely" joined the tour. Long's report of April 3, 1830, paraphrased company documents. Citing "the utility and importance of the canal in a national point of view," the report outlined a $40,000 channel-improvement plan.[46] In May, Congress added Back Creek to a lighthouse-improvement bill. The bill, however, fell to Jackson's pocket veto.[47]

Bernard and Totten played a much larger role in promoting what Americans called "the great national project," the Chesapeake and Ohio Canal. The Corps believed this canal across Appalachia would remake the capital city into a thriving crossroad, the vital link from the western rivers to the nation's largest bay. President Adams, comparing the project to the Seven Wonders, said the canal would one day be greater than the Great Wall of China. An early historian of the canal reached higher still. The Chesapeake and Ohio, wrote George Washington Ward, was nothing less than "the greatest public work which up to that time [had] engaged the attention of men."[48]

Actually the Chesapeake and Ohio, a fiasco, was a case study of bloated expectations. A stinging defeat for the Corps and Clay's American System, the waterway eventually did carry some boats but never repaid its investors. It never even lived up to its name. The project died a protracted death far short of the Ohio River, and its eastern tidewater terminus, Georgetown in the District of Columbia, was still 108 miles from Chesapeake Bay. Yet the story of the hapless canal is worth retelling for what it reveals about the nation's commitment to science and the fickle role of the expert as a compass of the federal state.

The idea of a water passage across Appalachia had long been the talk of Virginia. George Washington, a frontier surveyor, launched the scheme with his realization that the headwaters of the Potomac and the Ohio were less than fifty miles apart. Jefferson and Madison agreed that "Nature" had "declared in favour of the Potowmack" as the gateway to the Ohio.[49] With Madison raising funds for the project, Virginia, in 1785, chartered a Potomac company to improve the river to Cumberland and then build a toll road to the brown Monongahela, a tributary of the Ohio. Washington, the first president of the stock corporation, began hunting for a good engineer. In the 1790s slave and free labor went to work clearing rapids above Georgetown. Where the river dropped seventy-six feet at Great Falls, Maryland, the com-

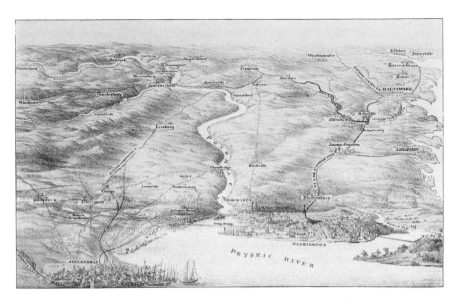

Route of the Chesapeake and Ohio Canal
from Georgetown to Cumberland, from
Edward Sachse's bird's-eye view of the
Potomac at Washington, around 1861.
Courtesy of the Library of Congress.

pany eventually completed a short canal with five wooden locks, but the in-
come from tolls was a trickle compared to the debt. Gallatin, Calhoun, and
Hemphill all said the project required and deserved substantial public assistance. Even Monroe backed federal aid. Charles Fenton Mercer, a congressman from Loudoun County in northern Virginia, a friend of Monroe's, saw
the canal as a chance to steal some of Baltimore's trade. Denying conflict of
interest, Mercer, in 1823, organized a new group of investors, making aid to
the Chesapeake and Ohio a condition of his support for federal internal improvements. Congress responded with $30,000 so that the Corps could
begin the surveys.[50]

The board for internal improvement immediately understood that the
Chesapeake and Ohio was the "most important" of its planning assign-
ments, a trial for Corps civil works.[51] In May 1824, the board sent out four
teams of surveyors: eight topographical officers, six civilians, five artillery
lieutenants, three fort engineers, and a West Point professor. Topographer
J. J. Abert led a party south and east from the Savage River in upper Virginia.
Captain McNeill scouted the headwaters of the Youghiogheny, a fork of the
Monongahela. A third brigade surveyed north from Pittsburgh to a point on

Lake Erie near Cleveland. A fourth examined a possible branch to Philadelphia by way of the Allegheny and Schuylkill, approximating the future course of the Pennsylvania canals. Although Totten considered wilderness surveying a "pleasant" occupation, clearly it was not.[52] Abert's party was hit by "the bilious fever common to the waters of the Potomac," probably yellow fever. Sweeping into the high country in the swampy heat of July, it lingered until "a regard for the lives of the party" forced Abert to suspend the survey. Calhoun ordered the surveyors back to the field, but the suffering persisted. "Disease," said Abert, was "our constant companion."[53]

The most challenging section was the highest elevation, the summit level, where the improvement board suggested a four-mile tunnel almost two thousand feet above sea level. Lack of water at the summit was the curse of a mountainous canal. Near Lebanon, Pennsylvania, the search for a summer supply had complicated construction of Canvass White's Union Canal. Congress pressed the engineers to determine with accuracy if the C. & O. might meet a similar fate somewhere in the Virginia highlands. When a brigade of academy graduates discovered a possible source of water at Deep Creek, a tributary of the Youghiogheny, Calhoun dropped the affairs of state to witness the test. There, in the driest season the engineers found a stream that could fill a 60′ × 12′ × 10′ canal lock in thirteen minutes. Stockholders rejoiced. Congress, greatly encouraged, endorsed the canal company's charter by an act of March 3, 1825.[54]

Little was said of expense until October 1826, when the board for internal improvement released a second enormous report. Minutely detailed and calculated to the fraction of a cent, the report was a high expression of hydrographical science, a clearly commercial application of the emerging West Point approach. Physics and trigonometry allowed the engineers to determine the most efficient shape of the canal bed—a trapezoid. Soil analysis projected ten types of excavation from quicksand to hard slate, each broken down to the probable cost per cubic yard. The report also staked out an economic position, a theory of public works. "When a nation undertakes a work of great public utility," the engineers explained,

> the revenue is not the essential object to take into consideration: its views are of a more elevated order—they are all, and, it may be said, exclusively, directed toward the great and general interests of the community. . . . When these national interests are satisfied, the principal object for which the work is undertaken is accomplished; and the fiscal advantages derived from the canal, and which would be an essential point to a company, becomes, in this case, of merely secondary importance for the nation.[55]

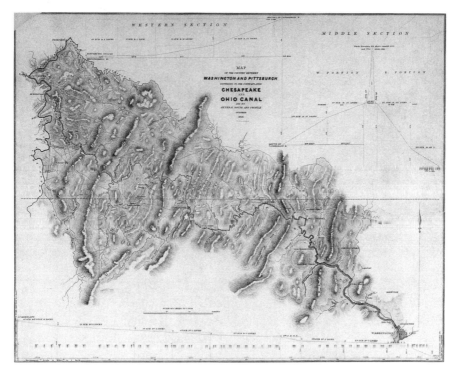

Corps map of the proposed Chesapeake
and Ohio Canal from Washington to
Pittsburgh, 1826. Courtesy of Record
Group 77, Cartographic Division,
National Archives.

Science and glory, in brief, outweighed fiscal concerns. Rejecting a shallow and narrow waterway, an Erie-type canal, the engineers wanted a large
excavation that would move heavy traffic, a canal six feet deep and sixty
feet wide with one-hundred-foot locks. The cost, said Bernard, would be
$22 million.

To assess the worth of $22 million is to imagine the prospect of selling
220,000 shares of stock in a market already depressed by a banking panic. In
1826 a good laborer, someone "able to excavate and throw in a wheelbarrow
15 cubic yards a day," might command ten cents an hour.[56] By these wages
the Virginia Board of Public Works had figured a canal could be cut for
$1.5 million. The entire 363 miles of the Erie had been built for just over
$7 million. But in the estimation of the U.S. engineers, a canal from Georgetown to Cumberland, the relatively flat first leg of construction, would cost

about $8 million. Investors welcomed this estimate like an ominous plague. Some abandoned the project. Others resorted to a desperate ploy: they called for a convention to discredit the army's estimate, and on December 9 the investors appealed to Congress to have the initial section resurveyed, not by academy graduates and certainly not by foreigners but by "practical civil engineers," men of long experience who had actually built canals.[57]

The confrontation was classic and as stark as any between scientific officers and self-educated civilian engineers. At one extreme stood Simon Bernard. As ranking officer of the improvement board, it was Bernard—assisted by Totten, Poussin, and the civil surveyor William Howard—who compiled, authored, and took responsibility for the Chesapeake and Ohio reports. A baron of the French empire, an aide-de-camp to Napoléon, and a brigadier general under Louis XVIII, Bernard was a patriot of both republics, as committed as Lafayette to the Franco-American alliance and the strength of the national state. He had been something of a prodigy in his hometown of Dôle. The son of an impoverished nobleman, Bernard had been educated by Jesuits at a charity school. In 1794 he had walked to Paris, nearly died of exposure, and at age sixteen, won a competitive berth at the newly organized Ecole polytechnique. As a young captain of engineers, he once had the temerity to tell Napoléon how to outwit the Austrians near Vienna during the Austerlitz campaign. Napoléon reprimanded Bernard but spared him from the upcoming battle and singled him out for promotion. His great work as a builder was the fortified harbor at Antwerp, Belgium, a tribute to the science of Vauban. His darkest hour was in Belgium as well, at Waterloo. After the French defeat, Bernard had hoped to accompany Napoléon into exile but was barred from the emperor's vessel, then banished from public life. Closely watched in Paris, his services no longer required, Bernard considered employment in the Dutch and Bavarian armies and rejected an offer from Russia. Lafayette suggested America. In the late summer or early fall of 1816 the engineer left Europe for a second career in the distant United States.[58]

Madison and Monroe welcomed Bernard and promised a large assignment: he would tour America, planning fortifications and advising the War Department on matters of administration and science. On November, 16, 1816, Bernard became a brigadier general and "assistant" to the president and the Corps. For all his fame and noble past, the engineer was a democrat who, said Napoléon, "would have preferred Washington to me." Monroe found him "a modest, unassuming man" whose appointment posed no threat to American builders.[59] General Swift thought otherwise. "It is humiliating," said the chief engineer before resigning his command; "a re-course to Foreign Engineers will in my opinion destroy the emulation of the American corps and finally ruin it."[60] To the extent that his background was largely foreign to American engineering, Bernard seemed a man of swollen stature, the army's aristocrat par excellence.[61]

At the profession's other extreme were James Geddes and Nathan Roberts, the two civilians appointed by President Adams to resurvey the canal. Both were Erie engineers; both of that class of self-made canalers, who, in the words of a later engineer, "knew by intuition what other men know by cal-culation."[62] Before his appointment as surveyor for the New York canals, Geddes had no technical experience. A lawyer by training, a judge and politi-cian by profession, he had used a carpenter's level only once, and then only for a few hours to settle a land dispute. Roberts was an upstate New Yorker of similar ilk. He was "a man of marked and peculiarly American character," his biographer recalled, meaning, perhaps, that he had risen from a sur-veyor's assistant to the height of his profession "with little aid from the pub-lished works of engineering."[63] This capacity for action—so unlike the meticulous French—had astonished the world when New Yorkers com-pleted the mountainous Erie Canal. Geddes and Roberts were proof that un-schooled sagacity could substitute for what officers called science. But the pragmatic New Yorkers failed to see that the rugged upper Potomac was not the Hudson or Mohawk. Its wilderness was steeper, its navigable streams were farther apart, and its sparse population was far less committed to a single public improvement.[64]

Predictably the civilians came up with statistics quite unlike the numbers used by the Corps. Where Bernard wanted a width of sixty feet, engineers Geddes and Roberts, reporting to Congress in 1827, said forty feet would suffice. Bernard advised lining the bed with stone and sloping it to facilitate the movement of boats according to the principles of hydrodynamics. Geddes and Roberts dismissed the science of water in motion. Citing "experience," the civilians claimed that simple, perpendicular walls were more durable and much cheaper to build. Moreover, said Geddes and Roberts, there was no need to line a hard-rock canal with additional stone. Wooden aqueducts and canal locks would further cut expenses, and thus, the civilians reported, the army had overestimated the cost of construction by at least 100 percent. Geddes and Roberts promised to complete the first leg of construction for half Bernard's estimate—about four million dollars.

Congress, anxious to build, rushed to support Geddes and Roberts. On February 11, 1828, Congressman Mercer lectured the House on the value of

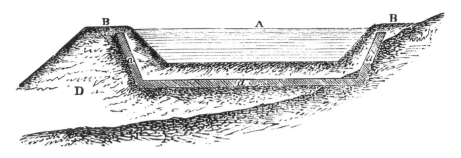

Cross-section of a sloped, stone-lined
canal bed, from Dennis H. Mahan, *An
Elementary Course of Civil Engineering,*
5th ed. (New York 1851). Courtesy of the
Office of History, U.S. Army Corps of
Engineers.

practical knowledge, making a clean distinction between "simple deductions" and the army's preposterous science.[65] Native instinct—as Mercer saw it—had defeated bookish abstractions. On May 24, Adams and Congress backed the Chesapeake and Ohio Canal Company with a one-million-dollar subscription of stock. Along the surveyor's line in Cumberland, Old Town, and Martinsburg it was an occasion for speeches and banquets. A cannon sounded in Georgetown. Washingtonians hoisted a flag. At a stockholders' meeting in June the old Potomac Company gladly surrendered its holdings to Mercer's new corporation. Politicians from Virginia, Maryland, Pennsylvania, and the District of Columbia sat on the board of directors. Mercer became president of the stock corporation. Erie Canal builder Benjamin Wright, the man who had recommended Geddes and Roberts, became the canal's principal engineer.[66]

Five years had passed since the Chesapeake and Ohio investors first convened in Washington, four years since Congress had authorized the preliminary surveys. By 1828 the delay had exacted its toll. Once the canal promoters had hoped the federal government might outright purchase the project. Now with the rise of Andrew Jackson the investors feared they were on their own. In four years, moreover, the advance of steam locomotion was shaking the conventional wisdom about the need for frontier canals. "Canals," said a New York rail promoter, "are but imperfectly calculated to answer the wants of the travelling public."[67] Virginians, however, still dreamed of a waterway across Appalachia, and no word of caution could sway the canal investors from the precipice looming ahead.

Suddenly in July 1828 the investors had cause for alarm. On Independence Day, just as President Adams was turning the first spade of earth for the Chesapeake and Ohio, a rival enterprise was breaking ground about forty miles away. It was the nation's first successful railroad, a municipal venture called the Baltimore and Ohio. Both the railroad and the canal began building along the Potomac. At Point of Rocks, a narrow gorge just below Harpers Ferry, both claimed the same piece of ground. The canal company sued and won. The railroad appealed. Courts deliberated while both companies lost time and markets to the Pennsylvania canals. At last, in January 1832, the Maryland Court of Appeals upheld the prior claim of the Chesapeake and Ohio, but the Maryland General Assembly insisted on a compromise that forced the canal to share the gorge. It was the beginning of the end for the great national project. Plodding ahead, the railroad reached the Ohio River in 1853. By then the canal company had already abandoned construction at Cumberland, some 30 miles short of the Appalachian coalfields and more than 150 miles from its objective in western Pennsylvania.[68]

Competition from Baltimore's railroad was by no means the only problem. Land disputes, labor riots, and a cholera epidemic kept the canal financing teetering on the brink in the 1830s and 1840s. There were also technical problems. The challenge of building next to the river, the steepness of the grade, a scarcity of mortar and stone—these and other obstacles rapidly deflated the optimism of 1828. The army did what it could to rescue the project. Federal troops tried to police the work camps during ethnic warfare and strikes. At the company's request in 1831, the army improvement board sent topographers Abert and Kearney on a canal-inspection tour. The board also suggested the tidal lock at the mouth of Rock Creek that kept water in Georgetown's harbor. Surveyor Howard mapped a Chesapeake and Ohio extension through Washington City and, working with Bernard, Abert, and others, projected a branch canal to Baltimore. In 1835 the topographer William Turnbull began the massive stone piers of the Potomac Aqueduct, an extension to Alexandria, Virginia, one of the largest civil-works projects of antebellum times. Army involvement continued in 1841 when the lighthouse builder William Swift investigated the canal for Congress. That year the company suspended construction while the investors scrambled for funds.[69]

In the end the great national project was a technological challenge well beyond the resources of Congress. A builder's nightmare, it vindicated the meticulous work of Bernard's engineers. By October 1850, according to company figures, the canal to Cumberland had cost about eleven million dollars—nearly three times the Geddes and Roberts estimate. In the end

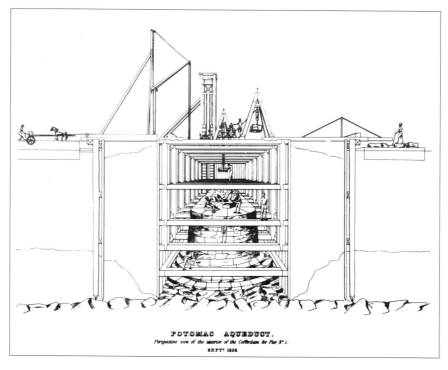

POTOMAC AQUEDUCT.
Perspective view of the interior of the Cofferdam for Pier N° 5.
SEPT: 1838.

Building the Potomac Aqueduct (a.k.a. Aqueduct Bridge), 1837. Cofferdams allowed crews to work at depths of 36 feet. Courtesy of Record Group 77, Cartographic Division, National Archives.

Below:
The Chesapeake and Ohio Canal at Georgetown, 1865, showing the army-built aqueduct over the Potomac. Courtesy of the Library of Congress.

even Bernard had underestimated expenses by more than two million dollars, but adjusting for the cost of the land and a quarter century of rising prices, his accuracy was uncanny.[70]

Near bankruptcy taught the canal company a lesson about practical engineering. Harder to grasp was something less obvious than accuracy and error, a sense in which competition between army and Erie engineers had been a contest of values and styles. "We were planning a work for the nation," wrote Bernard in his second report. "It did not belong to us to curtail the cost, in order to derive from the capital a greater interest . . . to the detriment of durability and conveyancy."[71] Bernard said permanence, efficiency, and stately construction outweighed fiscal concerns. But the New Yorkers had begun with quite different objectives. Geddes and Roberts thought less of the grandeur than of "the delicacy and importance of the trust reposed in them" by the friends of the venture. They proposed no greater monument to civil engineering than a waterway "of the same size, in all respects, [as] the Erie Canal," its aqueducts like those of the Mohawk, its stone and cement similar to "the production of the forests and fields of New York."[72]

If Geddes and Roberts were guilty of confusing Maryland with their native New York, then Bernard can be faulted as well. "He did the United States the honor to believe that Congress was perfectly serious in its intentions," wrote the historian Ward. Bernard wrongly assumed that the growing United States was "entirely competent from a financial point of view" and that Congress would spare no expense for a truly spectacular project.[73] Bernard thought of America, that is, in terms of France—a centralized republic capable of long-term investment in permanent facilities, a work for the ages. Mercer and the investors of 1828 had little choice but to reject Bernard's alien theory of canal economics. But to the extent that the company did follow the Frenchman's advice—enlarging the canal, insisting on the highest construction standards—Bernard almost assured that the great national project would never reach the Ohio.[74]

The canal builder himself was not in the United States to watch the story play out. As excavation near Georgetown was just getting under way, Charles X, the hated monarch of France, was overthrown by a liberal regime. In December 1831, Bernard sailed for Paris. Six months later he returned to the States and resigned. Ahead lay a turbulent third career as grand officer of the Legion of Honor, aide-de-camp to the king, inspector general of engineers, and minister of war. Bernard's aide Poussin also resigned for high office in Paris, leaving, for the first time since 1803, an all-American Corps.[75]

Without these talented planners the improvement board staggered and died. On June 22 1831 an order from the War Department transferred the survey program to Abert's topographical bureau. Abert quickly laid claim to the waterway program. Citing the august tradition of the Corps des Ponts et Chaussées, he argued the "clear truth" of the need for more West Pointers, stepping up the campaign to purge federal projects by phasing out civilian surveyors.[76]

Thus the board for internal improvement fell short of its great objective, but like the Corps and West Point, it transmitted an Old World ideal—the belief that science meant centralization, the conviction that planning was the province of scholars who served the militarized state. Bernard had anchored that Gallic tradition. Eager for grand assignments, he had stressed the false economy of temporary construction. A skilled bureaucrat, he had mastered the delicate art of weighing options and exploring alternatives only to recommend a single course of action in the final paragraphs of the report. Bernard was also the wise tactician who knew when to back down. In 1828, after Mercer had turned on the army, Bernard had stepped back from canals. Soon the engineers had retreated to the relative safety of river and harbor projects. Planning, the Corps had discovered, could be a way to find inner focus, the power to pause and assess what seemed to suit the organization and what clearly did not.

3. Expedient Methods, Frontier Machines

The laws of physics ensured that the federal water programs would survive and eventually prosper despite railroads and the squandered investment in the Chesapeake and Ohio Canal. Water was still the easiest way to move bulky freight: about twice as efficient as railroads, at least ten times more efficient than paved roads or common turnpikes. In 1839 a study by Charles Ellet claimed that railroads could haul a ton for about two and a half cents per mile. Canals could do the same for a penny less per mile, but steamboats on western rivers were much cheaper still. The Ohio-Mississippi system could transport a ton of cargo for as little as a half cent per mile. Fast and cost-effective, these broad interior rivers were the lifeline of the heartland, an interstate transport system, a tie from Pittsburgh to Cincinnati to St. Louis to New Orleans that emancipated commerce and broke the isolation of the trans-Appalachian West.[77]

Yet even in Ellet's time the hazards encountered by shippers were as great as the rivers themselves. Men still used a rope and a winch to "warp" a keel through shallow gravel. Pilots still tried to break through the passes below

The Mississippi River at St. Louis, 1853, showing a skiff tied to a snag and, at far right in the foreground, the end of a stone dike. Courtesy of the Missouri Historical Society.

New Orleans by ramming their ships into mud, unloading cargo, and waiting for the tide or a tow. Upriver were famous rapids—the Des Moines and Rock Island rapids above St. Louis, the eighty-five-foot drop in the Tennessee River at Muscle Shoals, the Deadman Island Ripples above Wheeling, and the furious falls of the Ohio. The Ohio-Mississippi system also had thousands of snags. These were the stumps and broken trees in the channel, the "planters" embedded in mud, the "sleepers" beneath the surface, the "sawyers" that swayed with the current, the "preachers" that bobbed up and down as if kneeling in prayer. Heavy watersoaked snags could easily impale a steamboat. In 1844, for example, the *Shepherdess* bound for St. Louis hit a snag in the dead of the night. A wheel flew off, the hurricane deck collapsed, and slowly the bow went under. Children froze in the river. More than 40 people died in all. In 1851 at least 120 perished when the *John Adams* struck a "sleeper" between Memphis and New Orleans. The boat split in half. A year later a Cincinnati newspaper reported that 57 percent of the serious accidents on the western rivers were boats lost to snags.[78]

Alexander Macomb, Chief Engineer, 1821–1828. Awarded a gold medal for heroism at Plattsburg in 1814, Macomb, swiftly promoted, left the Corps to become the commanding general of the army. Courtesy of the Office of History, U.S. Army Corps of Engineers.

Westerners were so keen on river improvements that some threatened to secede from the Union if the government would not do its part. When a second year of drought shut down the Ohio in 1819, a western navigation commission requested $10,000 from each of the six river states. Only Pennsylvania responded. Congress, meanwhile, had ordered the Corps to confirm what pilots already knew: that the Ohio had at least six bars of gravel and sand less than three feet in depth, that pilots on the Mississippi feared snags most of all, and that the channel above Baton Rouge was a tangled mass of driftwood. "The annual destruction of property, by these impediments, is immense," said Bernard and Totten, reporting to Calhoun and Congress on December 22, 1822.[79] The engineers thought that a system of river levees might control floods and expedite navigation. Machinery might be devised to pull snags from the channel. Dikes and dams that narrowed the river might deepen the bars. In late February 1824 these recommendations resurfaced in a bill taking shape in the House. Sponsored by Robert Pryor Henry, a Clay man from Hopkinsville, Kentucky, it called for $75,000 for snag removal and two or more Ohio River sand bar experiments. On May 11 the river bill passed the House by a wide margin, 115 to 60. Amended slightly by the Senate then signed by Monroe on May 24, the act put teeth in *Gibbons* v. *Ogden* by shifting the responsibility for steamboat navigation to the river experts who served the executive branch. Congressman Henry called it "the first attempt to test the great system of internal improvement."[80]

Monroe and Calhoun left the details of implementation to a resourceful bureaucrat, Chief Engineer Alexander Macomb. Colonel Macomb was already well known to Congress as the wily commander who, with help from the navy, had scared off the British at Plattsburg during the War of 1812. Like many of the early West Pointers, he went far on his family's name. His father, an Irishman, had been a tycoon of the Detroit fur trade. His mother was French. Macomb rose to the challenge of river improvement with something from both traditions: he sent Major Long to the Ohio to test a French-style wing dam; and, catching the mood of the West, he sponsored an unusual contest. The man with the best machine for pulling snags would get $1,000 and a chance at a federal contract.[81]

The snag-machine contest of 1824 brought out a parade of frontier inventions from the mundane to the truly bizarre. One man suggested a rock-filled "impulse boat" that would pick up speed with the current and jerk snags loose with a chain. Another said divers could saw underwater. A third wanted to dredge rivers by chaining a plow, like the tail of a kite, to a wide floating dam. The winning entry, however, was quite straightforward—a floating crane, simply a winch and a lever on two flatboats braced together about twelve feet apart. The inventor, John Bruce of Vanceburg, Kentucky, also won a $60,000 snagging contract. Macomb and Calhoun both had their doubts about Bruce, but the inventor, a campaigner for Clay, had powerful friends in the House. On October 12, 1824, Bruce pledged to remove "all trees, limbs and roots of trees, and logs . . . at least ten feet below extreme low water mark."[82] Macomb, having drafted the agreement, thought it "too clear to require explanation."[83] Events proved him wrong.

The main trouble with Macomb's agreement was that it tried to fix in a season what the rivers had wrought over time. Hundreds of snags had appeared every year as current cut into mud and trees lost their footing along the Mississippi's thickly forested banks. One pilot estimated fifty thousand dangerous snags on the Ohio-Mississippi system. Yet the man sent to oversee the machine-boat contract, Maj. Samuel Babcock, knew virtually nothing about western rivers. A fort engineer, he was a veteran of the late war with Britain and a builder tainted by scandal. Accused of incompetence at Fort Delaware, he was arraigned in 1824, tried, acquitted, and transferred to the remote Ohio. Near Pittsburgh he joined the contractor Bruce who was snagging his way downstream with four machines and thirty-two men.[84]

Babcock commended Bruce and showered the snag boat with praise. "Nothing remains which can endanger the navigation," Babcock reported to Macomb in October 1825; and, citing the "zeal" and "ingenuity" of the snagging operation, he added that the river was now safe for steamboats "in all stages of water." [85] It was a laughable statement and a public relations disaster. Stunned pilots, writing directly to Clay, claimed that the most dangerous snags were "untouched." [86] Bruce, in fact, had made no attempt to clear the Ohio for steamboats. Finding a loophole in the law and the contract, a phrase that referred to the trees that blocked navigation "at the lowest stage of the water," Bruce had threaded his way down the river, removing only the logs in the low-water channel during the driest months.[87] Soon pilots denounced Bruce as a trickster, and Major Babcock was roundly condemned as "one who knows nothing of the rivers Ohio and Mississippi, who has never navigated them, who knows not on which side the channel is." [88] Macomb suspended the contract. Bruce took the matter to court.

By 1826 the work of river improvement was employing more lawyers than engineers. Macomb, enraged, had Babcock charged and convicted for disobedience and gross neglect. President Adams had the man reinstated. Congress then settled with Bruce for an additional $6,200, yet the issue refused to die. As late as 1865 the Bruce-Babcock affair was still alive in the federal courts.[89]

The silver lining in the snagging debacle was that the Corps was learning a lesson: river engineering was still an uncertain science, and the army's superintendent, the man who supervised contracts, had to be savvy enough to mediate local disputes. In 1826 the popular choice was "the father of western steamboating," Henry Miller Shreve.[90] At age forty-one this stout, bullheaded pilot was already larger than life. An outspoken Democrat, self-made and largely self-taught, Shreve was to steam navigation what Daniel Boone had been to the Kentucky frontier—its mythic hero, its symbol of raw genius

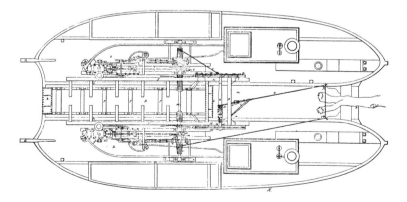

Henry Shreve's snag boat patent, 1838.
Courtesy of Record Group 77,
Cartographic Division, National Archives.

and frontier determination. Every riverman knew the story of how Shreve and inventor Daniel French had redesigned the engine and boat that took on and eventually smashed the Fulton-Livingston steamboat monopoly. Shreve was also one of the first to use a steamer in battle. In December 1814, while Jackson's army waited for munitions at New Orleans, Shreve had walled the steamboat *Enterprise* with bales of cotton and, under cannon fire, ran the British blockade. A contemporary called him "insane," but when the *Enterprise* pushed upriver to Pittsburgh, the event was widely acclaimed as a transport revolution, the dawn of the steamboat era.[91] A risk taker, Shreve had little patience for the army's statistical science. Yet even Major Long acknowledged that Shreve was "one of the most experienced navigators of the Ohio & Mississippi Rivers."[92] Macomb agreed, and under the provisions of the 1827 Rivers and Harbors Act the Corps created an office for western river improvement and handed the program to Shreve.[93]

Shreve, a civilian, was always prepared to jump the chain of command. When Macomb resisted his plan for a steamboat snagging machine, Shreve developed a model, wrote letters to high officials, relentlessly petitioned the House, and eventually struck a deal with Congress: the United States would front the money to build a prototype snag boat if Shreve would assume the financial risk. On April 29, 1829, the invention was ready for trial. An odd-looking machine with a strange-sounding name, the *Heliopolis*,[94] it was two boats joined by a platform with a windlass geared to the engine. Across the double bow was a board plated with iron, a "snag beam." Just what the snag beam was for was made bluntly clear at a bend in the Mississippi called Plum Point, Tennessee. Bearing down on the throttle, Shreve aimed the boat at a giant imbedded "planter," ramming the log head-on. A thunderous crash rattled the boat as the snag popped free in the channel. Then, before an astonished crowd, the machine pulled the log from the water so the crew could saw it on deck.[95]

That year the *Heliopolis* smashed and removed about two thousand snags. Crude and brutally effective, the boat was a river battering ram that owed nothing to West Point technique. "No machine can surpass it," admitted Lt. Alexander H. Bowman, the Corps' man in Kentucky.[96] In 1830 a Louisville committee of merchants and boosters petitioned for two additional snag boats. Shreve eventually built six at about $25,000 each. Westerners, grateful to Shreve and the army, compared the unusual boats to tooth-pulling dentists. They were "Uncle Sam's Toothpullers," symbols of federal power.[97]

Within five years of operation the *Heliopolis* had cleared the forested channel below St. Louis and new boats were at work on tributaries north and south—the Ohio, Cumberland, Arkansas, Missouri, and Red rivers. In

Logjam on the Red River. Courtesy of the
Office of History, U.S. Army Corps of
Engineers.

1832, according to Shreve, no steamboat on the Ohio River was lost to a
snag. During the next five years the number of steamboat arrivals at St. Louis
more than tripled while the rates for shipping insurance declined. So popular
was the snagging program that President Jackson permitted Congress to expand Shreve's operation. Jackson, in fact, approved about $3 million for snag
boats and channel improvements, more than that allocated by any other
antebellum president. One venture dear to the party in power was the work
in the Cumberland River from western Kentucky to Jackson's Nashville
plantation, a $100,000 project. Shreve, meanwhile, was straightening rivers
with dikes and using tin boxes of powder to blast through a seven-mile
stretch of the Mississippi above St. Louis called the Great Chain of Rocks.
He also declared war on the trees. Discounting the protest from pilots who
feared the danger of hidden stumps, the steamboat inventor maintained that
the best way to protect a channel was to clear-cut riparian forests before the
trees became snags. Shreve instructed his men to fell every trunk within
three hundred yards of the bank. During the height of the program, from
November 1842 to July 1845, government snag crews removed about eighty
thousand riverfront trees.[98]

The most famous and formidable battle in the war on the trees was Shreve's great success and his great frustration—the Great Raft of the Red. This snake-infested jungle had vexed transportation since tribes of migrant hunters had explored the knotted swampland in prehistoric times. Extending about 150 miles from Natchitoches to the Caddoan Indian camps of western Louisiana, the raft was a stagnant logjam, a tangle of vines and uprooted trees. Each year, as logs swept down the river, the mass grew by as much as a mile. In the 1820s the Corps had looked for ways to cut or bypass the raft, but two issues of the next decade forced a frontal assault. Planters from Kentucky and Tennessee were already using the Red to defy the Mexican ban against the import of slaves into Texas. In the early 1830s, moreover, the defeat of the Mississippi Choctaws touched off a rush for southwestern lands. Soon after the Corps released a study of the twisted river in 1831, Chief Engineer Charles Gratiot, Macomb's successor, said snag boats and an army of workers could dismantle the raft piece by piece. On July 3, 1832, Congress budgeted $24,000. Again the Corps turned to Shreve.[99]

Henry Shreve and his snag boat at the great raft of the Red River, about 1833. Copyright © 1970 The R. W. Norton Art Gallery, Shreveport, La. Used by permission.

Shreve arrived the following spring with a new snag boat, the *Archimedes,* three smaller boats, and about 150 men. After ten weeks the gangs had cut about seventy miles, yet Shreve reported to Gratiot that "the labour is severe."[100] Swarming insects drove men into the water. Storms unleashed violent floods. Whenever gangs tore out a section of logs, the channel would shift, whole islands would disappear, new trees would crash in the river, and Shreve would be forced to retrace his steps. Year after year for five seasons Shreve returned to the raft with a new allowance from Congress and up to 300 men. In the spring of 1838 the workers broke through at last. Steamboats pushed upriver as planters carved out plantations. A new city, Shreveport, was now America's springboard to Texas and the Indian lands of the vast Southwest.[101]

But the river paused only a moment to let Shreve savor his triumph. As a nationwide depression sapped internal improvements in the summer of 1838, cotton prices fell, banks collapsed, and a flash flood in Louisiana brought down a new mass of trees. The following year there was no help from Congress. With snagging and dredging stalled, Shreve went out on a limb to raise private money; but without guarantees from Congress the financing quickly went dry. In 1841 a new president and the chief engineer, both Whigs, lost patience with the founder of Shreveport. Marked as a Jackson man, the inventor was bluntly dismissed and replaced by another steamboating legend, John W. ("Roaring Jack") Russell, a Whig. Russell was a brawler more than a builder, a big man with a dangerous temper who had once towed a shack into the river and threatened to destroy a whole town because a thief had lifted a wallet. In 1842 the new superintendent began working with Long, Capt. John Sanders, and others to develop a river dredge and a new fleet of snag boats. Shreve fought back with a suit for patent violation. When the riverman died in 1851, the fight over Shreve's compensation was still unresolved.[102]

From 1827 to 1841 the Shreve regime in the West had been an era of the pilot-mechanic, a time when structural engineering—dams, dikes, and canals—gradually gave way to cheap but cost-effective frontier excavation. Soon a pattern emerged. West Pointers made ambitious proposals. Congress endorsed river projects but insisted on practical agents, typically local builders loyal to the party in power. There were constant financial setbacks, technical problems, or blind acts of nature—droughts, floods, heavy seas—that disrupted navigation improvement and sparked campaigns for reform. Over the years, however, mechanical snagging and dredging had opened new sections of river, and a host of new farms and businesses in waterfront communities had grown deeply dependent on federal aid.

Amphibious dredge boat, built
by Oliver Evans for Philadelphia's
board of health, 1804.
Courtesy of the Historical
Society of Pennsylvania.

OLIVER EVANS' CAR.

A good illustration of this boom-and-bust funding pattern was a waterway-dredging program that over committed Congress but diversified civil works. Dredging, like snagging, was a steam technology that owed more to Yankee invention than to methodical French science. The first steam dredge on record was also a steam-powered car. Built by Oliver Evans of Philadelphia and tested in 1804, the machine was an amphibious "scow," a bargelike carriage with a five-horse engine, a stern paddle wheel, and a belt of mechanical buckets to scoop out the mud.[103] Two years later the English inventor Samuel Bentham designed a "steam-dredging machine" for His Majesty's dockyards in London.[104] Baltimore, New Orleans, and other port cities conducted their own dredging experiments, mostly with hand-powered equipment. Federal contractors began digging out the Delaware River in 1803, but isolated expenditures followed by years of neglect did nothing to ease navigation. Not until the time of the general surveys, 1824 to 1838, did the Corps have the resources to study the long-term effects of marine excavation.[105]

The Corps sponsored no formal "contest" for dredging machines, but the rivers and harbors acts of 1824 inspired dozens of strange devices—human-powered treadmills, horse-powered capstans that dragged heavy scoops, grapples suspended from cranes, tubes filled with black powder to blast underwater, and sharp plowlike harrow scrapers that, when towed behind steamboats, stirred up the mud so the current could wash it downstream. Lt. George W. Long, Stephen's younger brother, invented a conical "diving bell" that lowered men underwater where they drilled and chiseled the rocks. Once rocks had been smashed into rubble, the Corps experimented with a spoonlike "dipper" on a pulley geared to an engine. In 1826 the engineers also spent $2,300 to develop a floating conveyor belt with rows or

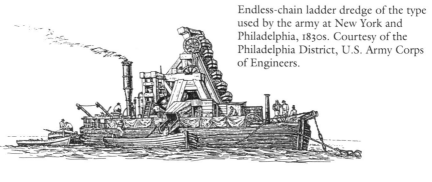

Endless-chain ladder dredge of the type used by the army at New York and Philadelphia, 1830s. Courtesy of the Philadelphia District, U.S. Army Corps of Engineers.

Cross-section of a ladder dredge, drawn by Corps engineer George Dutton, about 1830. Courtesy of Record Group 77, Cartographic Division, National Archives.

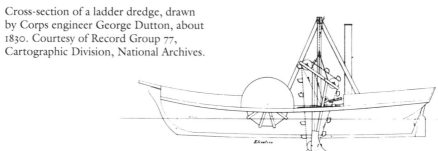

"ladders" of buckets on a continuous chain, a ladder dredge. On March 2, 1829, after the chief engineer reported success with ladder dredging in New York and North Carolina, Congress spent $139,000 for dredging at ten sites. Soon there were mud-clogged boilers and cracked engines from Lake Ontario to Mobile Bay. Rocks shattered cast-iron buckets. Chains snapped off in the water. In 1833 and 1836, steam dredges sunk off Savannah, blocking the harbor. Still a dredge boat was much less expensive than structures like dikes and wing dams that were used to excavate shoals by the force of current alone. By 1838, when Van Buren and the Twenty-fifth Congress began to dismantle the program, the Corps and the topographical bureau were deeply committed to dredging at about thirty sites.[106]

The main problem with mechanical excavation was that it had to be done every year. While the Corps was drawn to complex machines, dredging was often presented as an expedient maintenance procedure that might complement, not replace, scientific construction. Fort builders still wanted massive structures, such as breakwaters, dams, and canals. In 1829, for example, to clear and connect the maze of Florida rivers from the Atlantic to the Gulf, the board for internal improvement proposed dredging and the construction of a locked canal. Congress funded the dredging but dropped the canal. Likewise, in the blocked Delta below New Orleans, Capt. William H. Chase hoped to free the Mississippi with a $10 million canal. Cheaper solutions were "doubtful" and, at most, "temporary in their character," yet Congress decided to dredge.[107] In 1838 a Corps ladder dredge, the steamboat *Belize,* dug a nine-hundred-foot pass to a depth of sixteen feet. Next year the river shifted, the channel filled with silt, and the Corps' first attempt to open the Delta—a $285,000 investment—was entirely lost.[108]

The *Belize* affair, although a fiasco, did nothing to quiet demands for navigation improvements. Once a river had been snagged and dredged, once a port had been cleared for deep water shipping, the genie of army assistance refused to turn back. Waterway spending became a jigsaw puzzle of agreements with small pieces for dozens of sites. In 1835 a bill to continue the Delaware Breakwater included money for Shreve's fleet of snag boats. Next year the omnibus waterway bill split $476,000 among fifty disparate projects—a Connecticut seawall, a pier near St. Louis, eighteen dredging operations, a string of harbors along the Great Lakes. Sen. John P. King of Georgia, an Augusta Democrat, called the legislation a "monster." What King objected to most was the way Congress had added provisions until every state had its snout in the trough. When the Pennsylvania delegation backed a last-minute grant to Louisiana, King decried "this unnatural union between the French and the Friends," and the senator warned that the river lobby was "a current as irresistible as that of the Mississippi itself."[109] Yet King was a minority voice. On Independence Day, 1836, the waterway bill cleared the Senate by a twenty-one to seventeen vote.[110]

Piecemeal legislation frustrated army attempts to focus public resources on a few magnificent ventures. By 1836 the backlog of unfinished jobs exceeded one hundred projects, yet the official strength of the Corps remained fixed at twenty-two engineers.[111] The result was a growing reliance on about thirty officers of the line and up to fifteen civilians called "citizen" engineers. River and harbor assignments also gave mission and purpose to the agency selling itself as an executive department of transportation, the U.S. Topographical Bureau.

The rise of the topographical bureau was a challenge to Corps engineering. Smart and opportunistic, the agency thrived under J. J. Abert, the second chief of the bureau. Abert, highly respected, had studied engineering and law before enlisting as a topographer during the War of 1812. Like Roberdeau, Macomb, and Gratiot, he was a second-generation Frenchman, a stern man of science who delighted in arcane debate over what kind of builder in Europe could call himself a true engineer. "In Europe," Abert

John J. Abert, the long-reigning
chief of the bureau and Corps of
Topographical Engineers, 1829–1861.
Courtesy of the Office of History,
U.S. Army Corps of Engineers.

maintained, "the duties of a corps of topographical engineers are rarely extended to operations purely civil."[112] Yet the American situation forced
builders to mix traditions. Crossing from military engineering into civil surveying, a topographer became, in effect, "a member of civilian society"; and
there he would have to "conform" to the "dress, appearance, and living
exacted by the society into which he is thrown."[113]

Creative solutions to hazardous navigation were the topographer's bridge
between distant worlds. Major Long, the bureau's most experienced builder,
moved quickly from dams and canals to cost-effective expedients like snagging, ladder dredging, harrow scraping, and underwater excavation with his
brother's diving and blasting equipment. After a study of the Tennessee
River in 1832, Long soured support for canals by showing how stationary
winches ("warps") and tugboats could take barges through treacherous
rapids. He also came to regard navigational dams as "very questionable" because, Long told his biographer, "occasional and even frequent dredging
from the bars . . . prove more salutary and far less dangerous than wing
dams, jetties, &c., of any sort."[114] Others in the topographical bureau
turned more sharply away from dams and canals. Abert on the Kennebec
(1827), Kearney on the Tennessee (1829), Paul Perrault on the St. Marks
(1827), Napoléon Bufford at the Des Moines rapids (1829)—all saw alternatives to heavy construction in the frontier methods beyond *la technique*.[115]

As the bureau embraced innovation, it pushed for a break with the Corps.
One of Abert's first letters as chief was a petition for "a separation of duties
and command, and the establishing at Washington of a topographical bureau

under a topographical officer charged with the details of his own service." [116] The plan became War Department policy in 1831. Seven years later Congress took the logical step with legislation "to increase the present military establishment of the United States." Signed by President Van Buren on July 5, 1838, the act added twenty-one fort builders and twenty-six topographical engineers. Abert was promoted to colonel, and his bureau became an elevated civil-works and reconnaissance organization, a thirty-six-man Corps of Topographical Engineers. The new recruits were an amazing lot. Three in Abert's command had second careers in Congress. One served briefly as the governor of Maryland. Two became railroad presidents, and several earned fame as explorers. At least ten of the topographical engineers live on as place-names on maps of the West. [117]

Yet the Congress that built up the army also took something away. Responding to a bitter petition, James Buchanan of Pennsylvania, a Senate Democrat, said army engineers were banking "large fortunes" while neglecting their government work. [118] In 1838 the petition prompted a ban on the private employment of army officers. The new law, passed as an amendment to the army reorganization bill, repealed a vital provision of the 1824 General Survey Act. Some politicians still took the position that officers on furlough could help cities and states, but the one point never in doubt was that Congress had snapped a Hamiltonian link from the army to large corporations. [119]

The year 1838 ended a free-spending era. Historian Carter Goodrich and others have offered some good explanations—the wedge of the slavery issue that split the Whig coalition, driving the North from the South; the Seminole Indian uprising that called West Pointers to combat; the banking panic of 1837; the small-government ideology of Martin Van Buren; the advent of railroads and many new kinds of boats that drained support for canals. Another explanation was the growing sense among voters that the army's commitment to planning had trod on community rights. "If there is any thing which amongst our people is regarded as peculiarly odious," said Peter V. Daniel, a Virginia Democrat, "it is such a measure as implies in its execution or claims by its terms, an interference with the *territory of a State*." [120] In 1839 the Democrats killed the annual omnibus appropriation for rivers and harbors. In 1840, the nadir of federal spending, the budget for water projects was less than $2,000. [121]

Elder statesman John Quincy Adams was quick to pronounce federalism's early demise. "With me fell," he feared, "never to rise again in my day, the system of internal improvements by national energies." [122] What fell, in fact, was Gallatin's spectacular vision—a network of federal projects, grand objects of national pride. What emerged in its place were battle-wise army bureaus with an unshaken faith in the costly programs that strengthened the nation state.

4. Building Bureaucracy, Channeling Power

Long ago the sociologist Max Weber understood that bureaucracy's ability to control information was the ultimate source of its power. "The power position of a fully developed bureaucracy is always overtowering," Weber maintained.[123] When an agency brought in technicians to draft legislation and implement law, the details of policy-making dumbfounded elected officials. Government became a mismatch between experts and dazed dilettantes.

The federal water bureaucracy, in its infancy, was never independent enough to defy politicians outright, but the public-works committees in Congress knew very little about rivers and harbors. Confusion and political gridlock allowed experts to step forward with plans. In theory the planners were merely technicians who confronted great problems facing the nation by breaking down complex assignments into manageable tasks. In reality, however, the engineer-planning bureaus pioneered a scientific approach that left Congress with grand expectations. Weighing the cost of construction against benefits to the nation, the bureaucracy guided internal improvement through powers of implementation—the power to vie for jurisdiction through relentless self-promotion, to cultivate local support through regional superintendents, to assist corporations and sell the needs of commerce as concern for the nation's defense. Through these covert powers and others, public-works planning survived.

Today the influence of bureaucracy is a focus of political science, and yet, in a vast literature on the history of Corps divisions and districts, very little is said about politics or policy-making. Few historians venture beyond the thinking of Forest Hill, who, in 1957, said the Corps was "subservient" and divorced from politics by a keen sense of professional ethics.[124] Hill's important book anticipated a body of scholarly writing on the limits of expert knowledge. In 1962, for example, Richard Hofstadter referred to a "militant popular anti-intellectualism" that barred men with credentials from power.[125] Meanwhile two pathbreaking books by Leonard White had traced the rise of the executive branch without a word on the army planners who strived to consolidate power. White's bureaucrats were much like Hill's engineers: cautious, practical men with no theory of public administration beyond common sense, simple accounting, and respect for the letter of the law. Although a few determined generalists might win a congressional hearing—some cite the 1808 Gallatin report as a product of expert knowledge—government seemed to ignore the army's meticulous science.[126]

Hill and White seemed to be saying that the critical decisions in government were influenced by ideology and raw politics more than information—a valid point. But why in our cynical age have Corps historians failed to see what Corps critics have always contended: that professional values shape policy making, that engineers play a political role as brokers of federal power? Perhaps it is because of West Point. At the military academy, where the ideology of science is an unthinkable concept, there are few problems too complex for engineering solutions, and the army tactician, an empiricist, remains the impartial technician who lets facts speak for themselves. West Pointers, even today, strive for a neutral perspective by avoiding political discourse. "I learned to be an observer in politics, not an advocate," said Gen. Alexander M. Haig, Jr., a West Pointer. "Politics wasn't to be discussed by young officers in the dining hall. That was bad manners. . . . I made value judgments and kept them to myself." [127] Haig and his fellow West Pointers took pride in doing exactly what they were told. Dutiful service, however, requires a clear sense of direction, and seldom in the history of strife over army internal improvements have the president and Congress agreed to speak to the Corps with one voice.

Thus public-works engineering was political by its nature and politicized from the start. President Jefferson, after all, had first used the Corps and West Point to purge the Federalist army, and soon the appointment of army cadets was a guarded privilege of power. Artillerist George C. Strong, an applicant from a Democratic district in 1853, called the West Point appointment process "an elementary exercise in wire-pulling." Strong had heard that "an officer of the army has nothing to do with politics." Yet the young man suspected that family connections could be either a liability or an asset. "I had been taught to believe that considerations of party would overrule any merits," Strong recalled in his memoirs. His cynicism was confirmed when Congressman John Z. Goodrich, a Whig, rejected Strong's application. Goodrich "sacrificed me," said Strong. If a Whig were to appoint a Democrat, then, the soldier explained, "his [the politician's] official existence would certainly and abruptly terminate at the next election." [128]

Politics, then, was pervasive whether the army liked it or not. The fact that the Whig party won only two presidential elections meant that the party of Jackson and Polk was always well represented, and yet, on the question of internal improvements, it was hard for the army builders not to like what the

Whigs had to say. Abert, Thayer, McNeill, and J. G. Swift all favored the American Whigs. Long, often transparent, named his son Henry Clay. There were also engineer Democrats who weighed internal improvements against other burning concerns—engineers like John Sanders, the son-in-law of a Democratic congressman; Pierre G. T. Beauregard, a Creole aristocrat; and George Hughes, a Maryland secessionist. While some, like Sanders perhaps, played politics to advantage, others were devoted to partisan camps despite the party in power. Chief Engineer Totten, for example, remained a pro-Union, probusiness Whig through four Democratic administrations. A friend of John Quincy Adams, he became a trusted advisor to Whigs Winfield Scott and Zachary Taylor. In 1852 the colonel used close ties to the Fillmore administration to lobby for a costly project, the stone and cast-iron aqueduct that brought fresh drinking water to Washington, D.C.[129]

Not only did builders sometimes break ranks to lobby for critical programs, but the day-to-day management of a civil-works organization blurred politics and policy-making. Abert survived Democratic retrenchment by selling off dredging equipment and converting money from an office account to maintenance funds. His annual reports, meanwhile, challenged Tyler and Polk with horror stories from shippers, scientific testimony, shipwreck statistics, and urgent petitions for aid. Abert also made the most of vague regulations. The word "snag," for example, had once been taken to mean almost any river debris, but topographers narrowed the definition, excluding the stumps and logs outside the dominant channel. Abert then made the claim that the army's new fleet of snag boats was about twice as cost-effective as Shreve's awkward machines. Although the logic rested on some dubious economics— Abert omitted, for example, the cost of removing stumps along crumbling banks—the argument held, and snagging remained the focus of river improvement until war disrupted the program in 1861. Ultimately, the power to quote statistics and package information shaped government as much as the shifting mandates of Congress and the ambiguous directives of law.[130]

Powers of implementation extended federalism but stirred a critique of the army that dogs public works to this day. Once the issue of internal improvements had been largely a states'-rights dispute. By 1838, however, the resistance to national projects was also an indictment of army planning, an attack on the builders themselves. As Clay and Calhoun retreated to defensive positions and critics denounced the cancerous growth of the executive branch, a new kind of American scapegoat, the military bureaucrat, haunted the army's attempt to bring science and standardization to the river and harbor programs of the post-Jacksonian age.

"A Privileged Order of the Very Worst Class"

The attack on Corps engineering opened with a shot at West Point. "There is not on the whole globe an establishment more monarchical, corrupt, and corrupting than this," said *The Military Academy, at West Point, Unmasked,* a pamphlet printed for Congress. Published in 1830 by "Americanus," a pseudonym for deposed West Point superintendent Alden Partridge, the essay denounced the "gentlemen soldiers" who "sit high in *authority*" while others became "the *drudges,* yea, the mere *packhorses* of military service." West Point spat on democratic tradition. Screening the common man from high rank in the regular army, the academy coddled the sons of the rich, fostering "a privileged order of the very worst class, a military aristocracy." [1]

An aristocracy? The word seemed shrill at a time when the regular army was only about four thousand men. Yet Partridge's attack hit a nerve. In Jacksonian America, where politicians talked about "two classes" of men—those with college degrees and those with "a small degree of good sense"—a groundswell of voters believed that the clever mechanic or farmer could enter a profession or run the government with little formal training. [2] He could also command an army. "A man who would make a good overseer, or a negro driver, is better qualified for our service, than one who had received a first rate military education," said Col. Zachary Taylor, a slaveholder himself. [3] Frontiersmen mostly agreed. Davy Crockett of Tennessee said West Point had created a pampered elite. Caleb Atwater of Ohio, a travel writer, supported the men of West Point, but he admitted that Americans out west often viewed the elite engineer as a military dandy, a bookish man with a vest, snow-white pantaloons, a feathered hat, and a dangling eyeglass. On November 26, 1833, the critique took an ominous turn when the Tennessee general assembly—soon joined by legisltures and militia companies in Ohio, Connecticut, Maine, and New Hampshire—condemned "this aristocratical institution" and vowed to abolish the school.[4]

West Pointers fought back with a flood of statistics, but the first investigation, a massive response to the Partridge pamphlet, was far from complete. Compiled by the Corps for Congress in March 1830, the study claimed to review the appointments and family connections of 2,053 cadets. Only 49 appointments had gone to the sons of governors, congressmen, and other highly placed public officials, yet there was no data on wealth or social standing because, the army explained, "no testimonials with regard to pecuniary circumstances are required from those who apply for admission."[5] Eventually the chief engineer learned to keep better numbers. An 1842 investigation said 25 percent of the cadets were the sons of "planters" or "farmers." No cadet admitted his parents were wealthy. Eight out of ten claimed to be struggling boys from "reduced circumstances." Thirty-nine cadets said their parents' wealth was "unknown."[6]

Still, it was common knowledge that the fortified school on the Hudson groomed bright sons of the rich. Clay, Van Buren, Attorney General William Wirt, Secretary of War John Eaton, the chair of the House Committee on Military Affairs, and a president of Harvard all had sons at West Point. General Jackson sent two nephews to the Thayer academy, and one, Andrew J. Donelson, became a fort engineer. The sons of army generals also were well represented. Robert E. Lee, the seventh son of Gen. Henry ("Light-Horse Harry") Lee, went directly to Calhoun's office in 1824, presenting himself to the secretary of war with letters and a petition from five U.S. senators and three congressmen. One letter implored Calhoun to consider "the services of the Father" when judging "the claims of the Son."[7] Likewise West Pointer Pierre G. T. Beauregard, another son of a prominent planter, had his father write to the governor, who wrote to a congressman, who reserved the young man a commission through the secretary of war. While Whig editorials defended West Point, and Colonel Totten insisted that "the sons of the wealthy and independent in life comprise but a very small proportion of the young men at the academy," others were forced to admit that "in most instances the selection for West Point will be made from the sons of those whose influence will be felt at elections."[8] A House probe of 1837 made the obvious point. "When the same privilege is offered to the rich and poor," a select committee reported, "the influence of wealth possesses a most decided advantage, and will, in most cases, bear off the prize."[9]

At the height of the status dispute came a damning critique of internal improvements. In February 1836, as Washington buzzed with talk of

The United States Military Academy from
an 1840s watercolor. Courtesy of Special
Collections, U.S. Military Academy.

featherbedding in the Post Office Department and General Land Office,
House Democrats published an investigation of "fallacious" statistical information that drew the government into controversial enterprises. "Unfortu-
nately for the public treasury," the House reported, "some accident has
interposed, some foundation stone in the edifice has been displaced, some
unexpected change in the current of surrounding waters has been detected,
or some pelting violence of wind or ice, or other resistless power has oc-
curred, and just in season to disappoint the long deferred hope of both
government and agents, and to cause, or threaten, the downfall of the structure."[10] Some said the Corps was in league with the contractors to profit at
the public's expense. On March 4, 1837, President Jackson left office with a
parting shot at "unconstitutional expenditure," "corrupt influence," and
unnamed "powerful interests."[11] A year later, Chief Engineer Charles Gratiot was cashiered from the army for fraud.

Some attacks on the Corps were products of egalitarian times. Some,
however, were timeless. Jackson's lowbrow campaign against President Adams

helped turn the fear of the army into a lasting distrust of federal expertise. Jacksonians promised a "salutary system of reformation into every branch of government," a "clean sweep."[12] As new men in powerful office perfected the antipower platform, the center of government became "a hateful place," and its soldiers, said a Kentucky politician, were an elite force of "Capitoline guards," puffed and effeminate "half civilian[s] . . . who were dancing attendance at the skirts of Congressmen, or basking in Executive sunshine; never seen in the hour of danger, and found only where favors were to be had."[13] Barbs thrown at the army also hit civil servants. The duties of public officials were "plain and simple," Jackson told Congress; and the president, thinking perhaps of the Post Office and Land Office as well as the War Department, was convinced that "more is lost by the long continuance of men in office than is generally gained by their experience."[14]

Historians of engineering have neglected these kinds of attacks, presumably because, in the words of Professor Hill, "most criticism was directed toward Congress."[15] Yet the vilification left a permanent mark. Conditioning the Corps to conflict, it taught the agency to hunker in the defensive position that sheltered traditional turf. In many ways the beleaguered organization remains a child of the conflict between an elitist army tradition and inborn democratic ideals.

1. Lines of Resistance

The Corps learned defensive tactics after an attack on one of its own. Col. Charles Gratiot, Van Buren's chief engineer, was a second-generation French American, a slight man with a bright expression, a soldier of breeding and culture who befriended young engineers. His sudden fall from power in 1838 sent a shock through the War Department. Dismissed for transferring army money into his own bank account, the colonel, said his accusers, was a devious public official with a hand in the federal till. To the Corps, however, the man was "pure and good," a "gallant soldier . . . cast down from one of the most honored places in the land," a victim of the campaign to scapegoat top engineers.[16]

His tangled story began with a payment made to his father—a frontier quid pro quo. St. Louis fur trader Charles Gratiot Sr., a French Huguenot, had been promised thirty thousand acres for his stand with the Continental Army during the War of Independence. Although the Gratiots never did claim the land, President Jefferson, seeking support from St. Louis in 1804, sent the eldest Gratiot boy to West Point to consummate the purchase of

Charles Gratiot, chief engineer, 1828–1838.
Courtesy of Special Collections, U.S.
Military Academy.

French Louisiana and reward the father's good deeds. The boy returned to the Missouri country in 1806 as a nineteen-year-old fort engineer. During the War of 1812 he defended Fort Meigs, survived typhus fever, built a log fort on Lake Huron, and in 1814, led a brilliant attack on a well-supplied British blockhouse. In 1819, after duty in Delaware and Michigan Territory, Gratiot took over the biggest American fort project of antebellum times— Fort Monroe at Hampton Roads, Virginia, the massive star-shaped defenses at the mouth of Chesapeake Bay. When Alexander Macomb stepped up to major general in 1828, Gratiot became chief engineer. Over the next decade he prepared budgets and managed million-dollar appropriations for large projects like the National Road and the Chesapeake and Ohio Canal.[17]

Gratiot's career spanned an era when loose regulations and easy access to federal money made the life of a public official a banquet of great temptations. In 1821, for example, the House accused the Corps of "singular neglect of duty" for granting a Hampton Roads stone contract to Elijah Mix, the brother-in-law and business partner of a War Department clerk.[18] Gratiot was not implicated, but the scandal touched Swift and Calhoun. As Congress tried to tighten the rules, the builders took advantage of extra rations, outside salary supplements, and laws that granted clerks an allowance for the disbursement of federal funds. Gratiot said he deserved a double allowance—a $2 per diem bonus for Fort Monroe and another daily bonus for disbursing federal money at a nearby project, Fort Calhoun. When an auditor disallowed the double allowance, Gratiot appealed, and while the matter was pending, he took the disputed sum—$5,758—from one of the army's accounts. For years the War Department looked the other way until tough words by President Jackson and a series of army scandals forced the auditors to clamp down. At first the Treasury Department tried to reclaim the money by stopping Gratiot's pay. The engineer retaliated by making some dubious personal investments with army money. In 1837 the investments went bad, and the colonel was slapped with a bill for $31,713. On December 4, 1838, Gratiot, charged with fraudulent "defalcation," was harshly dismissed and replaced by the next senior fort builder, Joseph Totten.[19]

Muted in the wage dispute was Van Buren's dark perception that men in high office were robbing the treasury blind. Although it may seem outrageous today, the private investment of public money was tolerated up to a point in the 1830s. Bank directors and military officials were sometimes allowed to borrow from public accounts, but by Van Buren's time the practice was gravely abused. One Treasury Department study showed that 2,760 disbursing officers had defaulted on loans made with government money by Jackson's last year in office. Two years later, just months before the Gratiot incident, a New York customs official and a federal attorney had escaped to London with about $1.4 million. High salaries were another explosive issue. Gratiot, with two servants, a food allowance, and an annual income of about $3,900 a year, was making almost as much as any engineer in the country. In 1835 the railroads paid chief engineers about $3,000. A rail worker was lucky to earn $0.75 a day.[20]

In the end, after eight years of litigation, Gratiot could not escape the fact that he owed the government money, although the courts reduced the amount. A final petition in 1852 helped the colonel save face. At the urging of General Scott and other Whig politicians, the Senate Judiciary Committee agreed that Van Buren had acted rashly and that Gratiot may have had a "reasonable" claim.[21] Three years later, at age sixty-nine, one of the first army builders to handle millions of dollars died an impoverished man.

Although the Whigs dismissed the dispute as a cruel misunderstanding, the Gratiot affair was part of a larger story: a fight over the administrative jurisdiction of the engineering bureaucracy and a dread of technicians in power. By 1838 that anti-Corps resistance had three durable themes. First was a deep anxiety in the Yankee psyche that still feared sedition from France. Writing home from the military academy in 1830, Cadet Henry Clay, Jr., the senator's son, referred to the West Point climate of "suspicion & aversion," blaming it on the fact that Thayer was becoming an outcast. "It should be remembered," the cadet told his father, "that [Thayer] belongs to the French school of Politicians."[22]

A second line of attack tainted the Corps as an agent of centralization. In an age when the two top American generals, Taylor and Scott, were both outspoken Whigs,[23] it was surely no coincidence that Gratiot's reputation rose, fell, and rebounded with the level of Whiggish aid to army internal improvements. On August 31, 1852, one day after a Whig president signed the largest waterway spending act of antebellum times, Congress used the Gratiot case to tweak tightfisted Van Buren.

A third line of attack was a challenge to Corps engineering, a technological critique. Port programs, especially, drew fire from populist critics. Why, Congress asked, were Corps operations at Oswego, Erie, Cleveland, and Chicago chronically over budget? Why was America's largest breakwater ruining Delaware Bay? How had the promise of West Point planning gone so sharply awry? When the Democrats summoned the votes to pose these difficult questions, the Corps was publicly censured. Soon the waterway lobby unraveled and the port assignments had passed to Abert's topographical bureau.

The project that best exemplified the trouble with port engineering was the doomed attempt to arrest Philadelphia's falling position as the hub of the maritime trade. Geographically, the Quaker City was at a serious disadvantage. While Baltimore and New York enjoyed deep all-weather harbors, Philadelphia, dependent on the Delaware River, was shut down from December to late February by thick sheets of ice. Each year dozens of ships were crushed in the frozen river or forced out to sea. Floating ice routinely cut anchor cables. It crippled the cotton trade and took Philadelphia out of late winter seed exports. "Very many of our most valuable ships are annually either ordered to return to ports more accessible . . . or . . . bear away frequently in distress," said the Philadelphia Chamber of Commerce.[24] Meanwhile the Corps was calling for timber and rubble "ice piers" at four sites in the Delaware River from New Castle to Chester. Simon Bernard also drew plans for a fortified breakwater with a lighthouse and a stone "ice breaker" at Cape Henlopen on the south lip of Delaware Bay. Bernard believed a fine harbor could be built in a few years for $222,500. But when the last stone was in place more than seventy years later, the Delaware Breakwater, already obsolete, had cost about $3 million.[25]

Steady funding began in 1828 with $250,000. Too massive for any one organization, the breakwater "harbor of refuge" was conceived by the Corps, managed chiefly by the quartermaster general, and built by an army of civilian contractors under the loose direction of architect William Strickland, a Latrobe protégé. Bernard's original plan called for a replica of the French breakwater at Cherbourg—a great wall of stones piled into the surf like a

Skating the frozen Delaware River at
Philadelphia during the coldest winter on
record, 1856. Courtesy of the Historical
Society of Pennsylvania.

long, squat pyramid, its ocean face slanted to deflect violent swells. The challenge was to block ice and the crashing ocean yet permit the passage of useful currents that flushed mud and sand. Delaware Bay, however, refused to
follow the plan. Instead of scouring the harbor, the disrupted tidal action
brought in new deposits, and in 1834 the Corps reported the rapid approach
of sandbars as shallow as three feet. Upriver the Corps' ice shelter at New
Castle was having the same effect. By 1839, when the funding expired, the
gateway to America's second most populous city was shoaling and filling
with sand.[26]

Initially Strickland took most of the blame for the trouble in Delaware
Bay. A founder of the American Institute of Architects who had studied
briefly in Britain, Strickland, although eminently qualified, never worked
well with the army. Q.M. Gen. Thomas S. Jesup considered the architect
a deadweight political appointment whose salary was an annual waste of
$3,500. Because Strickland, a Philadelphian, refused to live at the site, the

army was forced to keep watch on the project through artillery lieutenants and topographical engineers. Army inspectors numbered and weighed each piece of stone. An army disbursing officer prepared budgets and helped direct traffic in the busy harbor. The army quartermaster also employed a surgeon, a naval officer, mooring officers, deck hands, and a large crew of stone workers—more than 110 men in all.[27]

When the Democrats killed the general appropriation for harbors in 1839, Strickland had little choice but to pronounce the project complete. Already a $1.8 million investment, it was three separate structures—a half-mile breakwater facing the ocean; a shorter, thicker icebreaker angled north toward the Delaware River; and a rubble pier. Sandbars were still a hazard but, for shallow-draft schooners at least, the breakwater was adequate shelter. Each winter thousands of ships crowded into the harbor. More than three hundred masts could be counted during a single gale.

Yet the wreckage continued and complications remained. In 1853 a Corps engineer conceded the "great hazard" of shoaling and ice still blocking the bay.[28] One dangerous spot was a gap between the breakwater and icebreaker where the wind whipped through the harbor. When the Corps said it could plug the gap for about $900,000, the War Department, in 1856, cut the request back by two-thirds. Even then a penurious Senate rejected the plan. Engineers made do with $30,000 for minor repairs until the boom years for harbor improvements after the Civil War. During the 1870s and 1880s the Corps added height, extended length, faced the structure with stone, dredged, and replaced a rubble jetty with a state-of-the-art iron pier. At last in 1896 the breakwater and icebreaker were a single wall almost a mile long and about fourteen feet above the lowest tide. Bigger than any harbor project in the Americas and more massive than the giant breakwater at Plymouth, England, it was truly one of the world's most magnificent dikes. For all its size, however, the shelter was too silted and shallow for transatlantic freighters. In 1897 the Corps was again stacking stone for a new harbor of refuge just a mile from the original site.[29]

Big, expensive, and environmentally jinxed from the start, the Delaware Breakwater was the tempestuous future of army harbor construction. Politically, it signaled the rise of the pork-barrel "iron-triangle" alliances that still bind water agencies to shipping organizations and powerful committees in Congress. Technologically, the project showed that breakwaters were not simply oceanproof forts, that every stone placed underwater changed the current and sedimentation of a delicate ecosystem. Delaware Bay was proof that West Pointers knew very little about what to expect from the seacoast. "A good theory of waves," Mahan's textbook admitted, "is still a desideratum

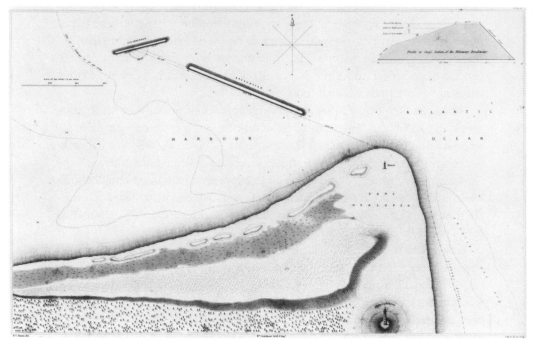

William Strickland's map and profile of the
Delaware Breakwater at Cape Henlopen,
1838. Courtesy of Record Group 77,
Cartographic Division, National Archives.

in science."[30] The Delaware Breakwater was half complete before Americans seriously studied the silt and sandy deposits that moved through harbors and bays.

Meanwhile the imperfect state of breakwater science made the Corps seem inept, even corrupt. Citing "extravagance" and "listlessness" in program implementation, Rep. Francis O. Smith of Maine, a Jacksonian, spearheaded the early campaign for engineering reform. On February 10, 1836, Smith damned the Corps with Gratiot's own list of troublesome water projects. Near the top was Oswego, New York, where the Corps had planned timber piers, a $33,000 proposal. Each summer for eight years the engineers anticipated success "beyond all doubt," each time asking for money. When Congress finally balked in 1836, the Oswego piers, still unfinished, were $93,055 over the original budget. Meanwhile a wing dam in Ohio collapsed. A Florida dredge-boat contractor was fired for negligence. A jetty in the Genesee River opened one channel but silted another. Piers at Dunkirk, New

York, allegedly completed "in a substantial and durable manner," could not survive the ice. "In none of these works," said Smith, "has the original estimate of cost, or of the probable effects of each expenditure, been verified by experience."[31] A House resolution calling for more precise estimates was followed by a heated Senate debate over army furloughs and the private employment of the topographical engineers.[32]

If a single project captured the mounting frustration, it was the deep bay at Erie, Pennsylvania. A grain emporium and a stop in the fur trade, it was reputedly "the best natural harbor on the Great Lakes."[33] Sailors still called the site by its French name, Presque Isle. Soldiers knew it as the thriving industrial town that had built a frontier navy for Capt. Oliver Hazard Perry during the War of 1812. In early August 1813, as the British squadron prepared for a beach invasion, Perry's men, toiling at ropes and poles, had managed to drag two heavy brigs across the soft four-foot bar that obstructed the mouth of the bay. Had either American brig not made it over the bar the battle of Lake Erie, a decisive victory for Perry, may have gone the other way. Soon after the war, the townspeople lobbied hard for federal money to remove the sandbar. Pennsylvania financed a bar survey and Congress, in 1824, added $20,000 for experimental piers.[34]

Little did Congress realize that small Presque Isle would require a full-time supervisor, a clerk, four carpenters, up to twenty laborers, and an investment close to $200,000 by the time the army suspended work in 1861. Initially the experiment fell to Corps engineer Theodore W. Maurice, a captain. A Corps history lauds Maurice as an "excellent example of the early Army Engineer," a man of "pragmatism and vision."[35] Yet Maurice, a fort builder, came to Lake Erie under the dark cloud of the Elijah Mix contracting scandal (the captain had given a stone contract to his father's employer), and the engineer's pier experiment, hardly pragmatic, was blind to the destructive action of tides in the delicate bay. The plan called for a double row of stone-filled cribbing to funnel the currents and tides that washed sand from the shipping channel. Completed in 1826, the project, said Maurice, "withstands . . . all the storms from the Lake and the Bay," but a storm a year later carried off most of the pier.[36] After a third appropriation in 1827, the chief engineer confirmed that an eighty-yard pier extension—another $8,000 investment—would cut through the sandbar. But the following year the Corps reported just nine feet of water at the entrance to Presque Isle. Deposits remained.[37]

Meanwhile a terrible gale had severed a thin neck of land at the opposite end of the bay. Tides swept in from the west, eroding the beach. Damming failed and so did a hastily constructed breakwater along the outer beach. The

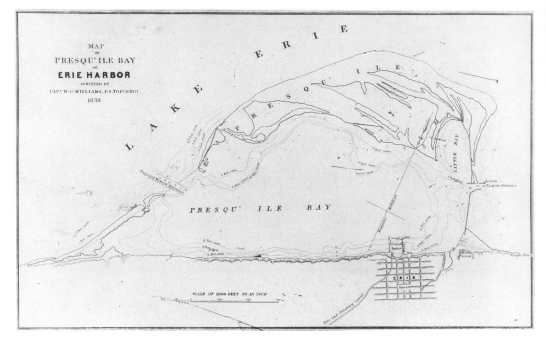

Map of Erie Harbor, 1839, showing the
eastern piers and the unexpected "western
entrance" breach. Courtesy of Record
Group 77, Cartographic Division,
National Archives.

bay now had two entrances, the second more navigable than the first. Soon
the builders were hit with a frightening realization: a large barlike deposit was
inching into the channel, filling the bay with sand. "It would be useless now
to attempt to estimate the ultimate expense," Totten told the chief engineer
in 1833.[38] Gratiot, nevertheless, requested $3,000 for another set of piers.

For years the engineers made no connection between shoaling, the breach,
and the works at the eastern end of the isle. So foreign was the study of tides
that the 1838 edition of Mahan's textbook cited only a single essay on waves,
and that was in French. But in 1835 a veteran snag engineer, Lt. Thomas S.
Brown, had the skill and detachment to see the cause and effect. Brown confirmed that the timber piers had cut through the bar as promised. There was
now about twenty feet of water at the entrance to Presque Isle. What was
beneficial at one point, however, was disastrous at another, for the tide that
washed through the channel changed the swirl of the current. Erosion now
threatened to close the safest harbor of refuge from Buffalo to Green Bay.[39]

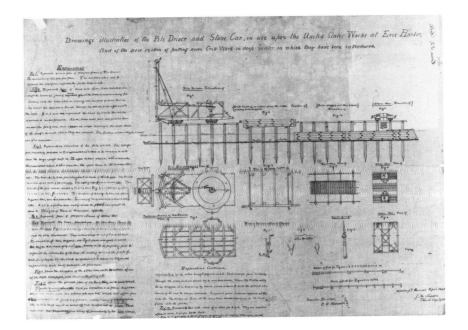

Drawing of a pile driver and stone car
used at Erie Harbor, 1845. Courtesy of
Record Group 77, Cartographic Division,
National Archives.

In 1836, 1841, 1852, and as late as 1902 the citizens of Erie continued to petition for federal assistance on the premise that the government's pier experiment was responsible for the ruin of a pristine bay. "The north pier has
already suffered several breaches," reported a town council in December
1841; "and there is every reason to apprehend, unless your honorable bodies
of Congress should provide for immediate repairs, that the greater part, if
not the whole of this work, will, very soon, be carried away by the severe
gales."[40] Soon the prophecy was fulfilled by a series of terrible winters. In
1843 the swells hit Oswego with such fury that a stone weighing fifteen hundred pounds was tossed onto a dock. Big Sodus Bay lost about three hundred feet of cribbing. Deposits re-formed at the mouth of the Cuyahoga
River in Cleveland harbor. Chicago was choked off by sand. As the western
ports cried for relief in the early 1840s, Capt. Thomas Jefferson Cram, a Lake
Michigan harbor superintendent, called the army's construction program "a
total loss."[41] The Chicago *Evening Journal* said Great Lakes sandbars and

inadequate beacons were responsible for more than thirty shipwrecks and groundings each year.[42]

Engineers faulted the bedlam in Congress, but the Jacksonians pointed the finger at the Corps. One critical vote of no confidence came on August 1, 1838, when an order from the War Department transferred harbors, rivers, and "all new works of [internal] improvement" to Abert's organization.[43] Whigs hoped that the topographical corps could save the American System. Democrats, however, recalled the sins of the past. Citing the "errors of calculation, want of economy, delays, and mismanagement" that had characterized the War Department's harbor program, the House Commerce Committee, reporting in 1842, said Congress should "long hesitate" before any more money was lost on public construction.[44] That year there was no general appropriation for harbors, and none followed until June 11, 1844. A modest investment, the 1844 act divided $485,000 among twenty-three western projects, all but two on the northern lakes.[45]

The one bright spot in the lake harbor program was the revival of army planning through a scientific survey. Created with $15,000 from Congress in 1841, the U.S. Lake Survey was a spark from the active mind of Lake Erie superintendent William G. Williams, a topographical engineer. Congress wanted harbor maps, but Williams, working through the topographical corps, proposed a precise hydrographic triangulation and an atlas of construction plans. It was the topographer's "earnest hope" that the lake reconnaissance would "serve as the basis of a great system" of improvements from Buffalo to Duluth.[46] Williams confirmed the need for a beacon and harbor of refuge about every hundred miles along America's interior coastline. He also gathered shipping statistics and made an exhaustive investigation of the passes between the lakes. At Sault Ste. Marie, where rapids blocked the channel from Lake Huron to Lake Superior, Williams proposed a six-mile canal. North of Detroit he called for a dredged channel through the St. Clair Flats. At Buffalo he wanted a long breakwater. When civil engineers attacked the elaborate breakwater proposal, Williams dismissed "the local prejudices of Buffalo" and, writing to Congress in 1842, defended his own design.[47]

These plans remained on paper during the 1840s. When President John Tyler vetoed a Whig bill to fund construction through a tariff, the former vice president, an enemy of Clay and the American System, said only the western projects deserved federal assistance. In 1845 the flip-flopping Tyler killed aid to the West as well. John Spencer, a prominent Whig, called the policy "incomprehensible."[48] As rivermen gathered in Memphis to protest, Abert sold equipment, cut expenses, laid off civilian river and harbor superintendents, and laced his budget proposals with urgent pleas for support.

Dock and loading facilities
on the Chicago River,
about 1855. Courtesy of
the Library of Congress.

At first the topographical corps had reason to be cautiously optimistic about the future of internal improvement under Tyler's dark-horse successor, President James K. Polk. A Tennessee Democrat, Polk had the blessing of Andrew Jackson, and Old Hickory had been kind to western rivers and harbors. But Polk not only vetoed harbor construction; he made the campaign against government projects a cause célèbre. On August 3, 1846, Polk killed the first of two large waterway bills with an attack on Congress and a condemnation of the "disreputable scramble" for federal aid.[49] Scoffing at western attempts to call every inlet a harbor, Polk said the Great Lake ports were removed from foreign commerce and therefore of minor importance. Westerners exploded with outrage. At the Chicago River and Harbor Convention of July 1847, more than ten thousand waterway boosters crowded into a tent pavilion to blast the president and remind Congress of its "obligation" to assist inland trade.[50] The growing sandbar at the mouth of the Chicago River was named Mount Polk.[51]

Inevitably the chief of the topographical corps was drawn into the funding dispute. Polk's secretary of war considered Abert "impartial,"[52] but the colonel, a Clay supporter, was a spirited nationalist, and his reports of the mid-1840s were read as a call to arms. "Without these harbors," Abert told the War Department, "the number of large towns which embellish the shores of these lakes would not have been built; the immense population upon their borders, and the extensive cultivation of such numerous tracts of land, would not have taken place."[53] Whigs claimed Abert's appeal gave a semiofficial sanction to federal harbor construction. Catching the Polk administration in an apparent contradiction, the House and later the Senate ordered reprints of Abert's reports. Speakers at the Chicago River and Harbor Convention also cited the topographical corps, and when the convention delegates petitioned Congress in 1847, they used Abert's statistics to attack Polk's position. Adamant, a defiant Polk said public works would interfere with the Mexican War. On December 15, 1847, the president struck down $60,000 for rivers and harbors in Wisconsin. Congress responded with a series of resolutions denouncing Polk and his veto power. There the matter rested until Democrat Lewis Cass, Whig Zachary Taylor, and Free Soiler Martin Van Buren turned the three-way election of 1848 into a national referendum on slavery in the territories and the power of Congress to fund public works.[54]

By Polk's last year in office the fight over harbor projects had battered the West Pointer's sense of himself as a detached man of science. The confrontation peaked at a time when the civil-works program was creaking under the weight of lingering scandals—the Gratiot litigation, a suit by Henry Shreve over his snag boat patent, a protest over the firing of Brooklyn dry dock superintendent William McNeill, a House investigation that lowered the pay of Mahan and another professor, and grumbling over the expenditure for coastal defenses.[55] The 1840s also revived the attack on West Point as an "effeminate" institution.[56] Pamphleteer Alden Partridge, searching for the ultimate insult in a long petition to Congress, said the despots at West Point were "military dandies."[57] Others called the West Pointers "delicate" and "effete."[58]

Stunned and seriously challenged, the scientific brain of the army had suffered damage, yet the nightmarish era ended as quickly as it had begun. The conquest of northern Mexico, a gold strike in California, the demand for a railroad to the Pacific, and a disastrous flood in Louisiana were four parts of a chain reaction that briefly renewed army construction. Soon the Whigs were back in the White House and a new round of water improvements had opened new channels of power.

2. Unflinching Professionalism

"Hip! Hip! Hurrah! War at last sure enough! Aint it glorious!" wrote engineer George B. McClellan, spoiling for combat.[59] McClellan could see that the Mexican War was a mismatch: a lean but mobile and "scientific" army against Santa Anna's slow-moving troops, modern artillery against obsolete cannons—a golden chance for the men of West Point. More than

450 academy graduates received honors and brevet promotions during that lopsided contest. Corps engineers Totten, Lee, Beauregard, John Lind Smith, and others became General Scott's field tacticians, his "Little Cabinet." Engineer troops called sappers and miners also showed promise in combat. Organized by Totten and fiercely led at the Battle of Contreras by engineer Gustavus W. Smith, the troops charged a bridge and routed an artillery stronghold on the San Antonio Road. Pushing north, Smith's men dogged the fleeing defenders to the gates of Mexico City.[60]

Nineteen topographical engineers also served in the war zone, but many remained less than elite in the eyes of the high command. Topographer Meade said he was permitted to work "as little as possible" while Totten's fort engineers hoarded the best assignments.[61] General Taylor, Meade complained, was unsure about how to use engineers. On the banks of the Rio Grande, where the "Army of Observation" prepared to launch the invasion, the lack of a bridge-building program cost Taylor nine vital days.[62] Still the topographical scouts found other ways to contribute. On September 8, 1847, Capt. Joseph E. Johnson, a former topographer, led 341 men into Mexican fire at the Battle of Molino del Rey. Ninety-eight Americans perished. Meanwhile Lieutenant Emory and three young West Pointers had mapped a path through the desert, the future route of the Southern Pacific. Hughes opened the road to Japala. Long built army steamboats in Texas. Frémont's topographical expedition joined the rebellion in California. Williams, the talented lake surveyor, was shot dead at the battle of Monterrey.[63]

A few West Pointers of conscience denounced America's conquest. Ulysses S. Grant, an infantry lieutenant, called the invasion naked aggression, but most saw the war as a triumph of science, proof that West Pointers could fight. "The Mexican War was our romance," said Simon Bolivar Buckner, Grant's close friend at West Point.[64] General Scott claimed that the skill and gallantry of "our graduated cadets" had been the decisive factor that reduced a difficult war to two brilliant campaigns.[65]

Applause for the fighting savants suspended the angry debate about what kind of American builder could best direct public works. Totten, bursting with energy, used the momentum of war to promote the Smithsonian Institution, standardize lighthouse construction, beef up the fort program, and test his ingenious design for iron artillery shutters that automatically shut after a gun was fired. Hughes surveyed a railroad across the Isthmus of Panama. Hard-fighting topographer Johnson, a wounded hero at Cerro Gordo, returned to arid Texas for a road and river survey. Abert, meanwhile, viewed the peace of 1848 as a springboard to new operations. Praising the tactical brilliance of Hughes, Emory, Johnson, and others, the chief called his topographical scouts "the eyes" of a frontier army. "Science," he insisted, "will ever predominate over numbers."[66] Boldly seizing the moment, Abert requested $45,000 to study the new domain, as well as money for marine hospitals, lighthouses, road and railroad surveying, and a sewer project in the District of Columbia.

An election year, 1848 also was a year of promise for the river and harbor crusade. While the Taylor campaign skirted the waterway issue, the Democrat Cass was widely denounced for failing to repudiate Polk. On May 26, Cleveland's *Daily Herald,* a Whig paper, reduced the Democratic platform to three controversial points: protracted war, the extension of slavery, and a cold shoulder to public projects. A month later an eloquent young Whig from Illinois, Congressman Abraham Lincoln, linked science to public improvement in a stirring address to the House. "The question of improvements is verging to a final crisis," Lincoln maintained, but how would Congress determine which projects were the most important? The solution was "statistical information," a stern reliance on facts.[67] Another voice for internal improvements was Millard Fillmore, the Whig runningmate. Placed on the Taylor ticket to placate Henry Clay, Fillmore, a Buffalo politician, helped the Whigs win the election by capturing upstate New York. When Old Zach died after a year and four months in office, the stage was set for a Whig revival, a second era of public works.

Fillmore had barely moved into the White House when a crisis in Louisiana forced the president's hand. A second year of serious flooding was turning about 27 million acres into an alluvial swamp. As scientists rushed to study the rising disaster, Congress, in September 1850, appropriated $50,000 for a Mississippi Delta survey. Soon the survey was mired in a professional contest. Although the work had been assigned to Abert's topographical corps, bridge builder Charles Ellet pressured the War Department to split the appropriation. Eventually the army funded two rival surveys. Ellet's report, the first submitted to Congress, was published in 1852, then expanded for a private printing the following year. A highly speculative treatise, it called for dams and storage reservoirs, a network of artificial lakes that would hold back high water and release it for navigation during the driest months. The scheme, a French idea, was foreign to American builders, and Ellet was quickly denounced. "Mr. Ellet's formula [for measuring river flow and flooding] is the worst formula ever proposed for its purpose," said Edwin Hale Abbot, an eminent engineer.[68] Riverman William Milnor Roberts recoiled at the huge expense of a storage-reservoir system, and Colonel Abert, although impressed by Ellet's imaginative presentation, claimed that high dams might have an "injurious effect" on the Delta.[69] More than a half

Andrew A. Humphreys—topographer, co-founder of the National Academy of Sciences, and the Corps' fiercely combative chief engineer, 1866–1879. Courtesy of the Office of History, U.S. Army Corps of Engineers.

century later the top Corps engineers remained skeptical about the use of reservoirs to control raging floods.[70]

Meanwhile, the army's other Mississippi Delta survey proved equally controversial. Authored by two West Pointers, topographers Humphreys and Abbot, it was a comprehensive hydrographical investigation, the first of its kind. The study pioneered new ways to measure the current, new theories to understand the mud bars that obstructed the Delta, and a new formula to explain the relationship between a river's "sediment load" and its propensity to flood. First printed in 1861, it quickly became a scientific sensation. *The American Journal of Science and Arts* called the Humphreys-Abbot report "one of the most profoundly scientific publications ever published by the U.S. Government," a "monument" to "unwearied industry and patient accuracy."[71]

Yet the massive Delta survey, although widely acclaimed, was also a blanket of caution that stifled some useful ideas. Critical and precise, it mirrored the training and bearing of the survey's principal author, Capt. Andrew Atkinson Humphreys of Philadelphia, a stout man with a frowning mustache. Humphreys made the most of the family connections that brought the rich to West Point. As the son and grandson of high-ranking navy officials, the young engineer grew close to the Bache family and Whig politicians like Secretary of War John Spencer—so close, in fact, that Spencer once intervened to have Humphreys relieved from duty in a dangerous swamp. Coast surveyor A. D. Bache also pulled strings for Humphreys. Assigned to the coast survey in 1844, Humphreys took over the Washington office while Bache was out in the field. In 1846 a letter from Bache to the War Department kept Humphreys out of the Mexican War, but when the topographer applied for the Delta investigation, Bache let him go. Order and diligence permeated the work in the Delta. "I cannot understand," Humphreys confessed, "how any man can be willing to assume charge of a work without making it his business to know everything about it from A to Izzard."[72] Lieutenant Abbot, Humphreys' young assistant, was awed by his coauthor's concern for the smallest details. That same microscopic vision made the topographers hard on builders like Ellet who proposed costly solutions and slighted army research.[73]

One idea rejected by Humphreys was the construction of Delta jetties that would funnel the sluggish water through walls of timber and stone. Jetties, most builders believed, increased the excavating power of the current. Flushing out silt and mud, they cut through the shallow deposits that made rivers back up and flood. Humphreys and Abbot admitted that jetties were sometimes effective—years before Humphreys had directed the Chicago River jetty project—but the Delta survey claimed that any increase in the power of the current would be checked at the mouth of the river by tides washing in from the Gulf. Humphreys and Abbot also had doubts about "cut-offs" through bends in the river and "outlets" that drained high water into overflow swamps. As for the Ellet reservoir scheme, the Delta survey stressed the "physical impossibility" of constructing dams high enough to contain seasonal flooding.[74] In the end the only sanctioned solution was a uniform system of levees that would reinforce the dominant channel and raise the height of the banks.[75]

For a time the weight of opinion was clearly with Humphreys and Abbot. After the war, however, the Delta survey came under fire, and the attack on General Humphreys, a symbol of the "privileged order," revived the dormant crusade against alien West Point techniques.

Professional disputes isolated the army but left a political plank where rivals could sometimes agree. Whatever the solution to floods, both Ellet and Humphreys believed that control of the western rivers was a task for federal science, a job too huge for the states. Even the aging Calhoun, an apostle of the states'-rights position on every issue but internal improvements, insisted that the great Mississippi was an "internal sea," a body of water with as much right to federal protection as Boston Harbor or Chesapeake Bay.[76] On December 2, 1850, Fillmore removed any threat of an executive veto in his first annual message. Echoing Webster and Clay, the president denounced the hair-splitting distinctions between "local" and "national" projects. All states had a common interest in safe navigation: the cotton ship grounded on a southern sandbar was probably owned in the North, and the proposed canal at Sault Ste. Marie, although only a mile, would extend American commerce more than one thousand miles. Fillmore encouraged the House to draft a spending agreement with projects for every state. Signed into law on August 30, 1852, the appropriation split about $2.25 million among one hundred improvements—sixty-nine harbor projects, twenty-three river assignments, two canal studies, snagging, dredging, and the U.S. Lake Survey. The next day Congress added $50,000 so that Humphreys could finish his work in the Delta. All together the Delta region won the largest appropriation, about $150,000.[77]

At first the War Department expected the two engineering organizations to spend the money together through a single board of internal improvement, but the rivalry between Totten and Abert scuttled that logical plan. On September 10, 1852, an order from the War Department created two oversight waterway boards—a board of fort engineers assigned to the coastal rivers and seaboard projects and a board of topographers to direct improvements out west. Each board had the authority to "examine, approve, modify, or reject every project or plan of civil improvement proposed by any officer or civil agent." Each, moreover, was responsible for "the quality of material and workmanship, and the general system of expenditure and administration."[78] Here was the seed of a far-reaching organization. Although the boards quickly disbanded after the funding dried up, the army's scientific discretion—its power to innovate, its authority to hire, fire, and shape policy through program implementation—was now formalized by regulation and confirmed by federal law.

Grand expectations amplified agency power. Because the 1852 rivers and harbors act authorized more construction than Congress was willing to fund, engineers pioneered a kind of cost-benefit analysis, a way of weighing options and assessing the relative worth of rival waterway schemes. In Maine, for example, a Corps lieutenant used commerce statistics to show that Rockland was more deserving than Owls Head, a neighboring port. At Milwaukee a topographer silenced support for small pier improvements by arguing the economy of a shipping channel. Elsewhere, however, the engineers had little choice but to return with empty pockets to Congress. Scare tactics were not uncommon. Topographer George H. Derby predicted the "utter destruction" of the fine port at San Diego unless Congress funded a levee.[79] At Chicago, where Congress had budgeted $20,000 for a $250,000 harbor construction project, Major Graham made do with dredging and waited for additional funds.[80]

Few engineers thought the decade would end without a new round of funding for water projects. During the fourteen years of the general surveys, 1824 to 1838, Congress had supported public construction with an annual allowance. But Franklin Pierce and James Buchanan, returning to Polk's hard-line position, rejected the notion that Congress could bind the Republic through a "general system" of public works. Pierce delivered the critical veto in 1854. Blocking money to finish what the army had already begun, the president admitted the need for a few public lighthouses, but he warned of "disastrous consequences" if Congress paralyzed the private sector or compromised the rights of the states.[81] Soon Pierce's secretary of war, West Pointer Jefferson Davis, was fighting Totten and Abert over small administrative details. Davis also shifted resources to the Office of Pacific Railroad Explorations and Surveys, a ward of the War Department. Launched with $150,000 from Congress in 1853, the railroad office captured some of the government's best surveyors.[82]

Vetoes and close supervision reined in the waterway builders, but a trickle of local funding and sheer bureaucratic momentum kept most projects alive. At Baltimore a cost-sharing agreement paid for dredging in Brewerton Channel. At Hell Gate in the East River a New York financier raised more than $6,000 so that Capt. William D. Fraser, a fort engineer, could blast and chisel a treacherous rock. On Lake Michigan, meanwhile, ports organized volunteer crews to repair piers and run army dredge boats. Graham supported the program through the sale of government sand.[83]

Army engineers also aided the states through waterway corporations. Although the Democrats had tried to block this kind of informal assistance back in Van Buren's time, there were exceptions. Port dredging companies and the Chesapeake and Ohio Canal often used army technicians. And, in 1857, when the Louisville and Portland Canal needed a new government loan, the company had Abert and Long present the case to Congress in a topographical survey. Another project pushed by the army was the canal at Sault Ste. Marie. Supported by private money, the state of Michigan, and a vast federal land grant, the Sault Canal, opened in 1855, was a semipublic "mixed enterprise" that used both government builders and Erie Canal personnel. The War Department contributed by allowing a veteran topographer, Capt. Augustus Canfield, to serve as the chief engineer. Canfield designed the working heart of the project—two masonry locks, each 350 feet long and 70 feet wide with a depth of 11.5 feet. Deep but not deep enough for big steamers on the upper lakes, the locks were soon obsolete. Yet the engineers remained active at Sault Ste. Marie. After the Civil War the giant Weitzel Lock with its gatelike movable dam made the busy canal a showcase for the latest construction techniques.[84]

Projects like Sault Ste. Marie showed that an agency determined to build could stay true to its waterway mission despite the state's-rights orientation of a cautious executive branch. In October 1852, when the governor of Michigan had requested the loan of an army surveyor to study the proposed Sault Canal, the War Department refused. But Canfield began a survey without official approval. He even offered to work without pay. During the Pierce administration, four years of legal debate over federal interference, help from zealous West Pointers was applauded and elsewhere denounced. In 1854 the *Detroit Free Press* called Canfield a "remarkable inventive genius."[85] Secretary Davis, however, frowned on informal assistance and tried to have Canfield transferred. Soon a California senator and civil engineers in Philadelphia, New Orleans, and New York were again asking Congress to ban army assistance. In 1855 the railroad engineer Herman Haupt referred to "the evils of public management" that turned men into pompous machines.[86]

Both faces of public construction—the industry, the arrogance—were evident in the District of Columbia, a gleaming city for Congress, a laboratory for West Point technique. Army engineers had directed public works in the city since Washington had hired L'Enfant to lay out the streets and parks. Engineer officers also helped survey and build the city's sewer system, the Potomac Aqueduct to Alexandria, the harbor at Georgetown, the brick Smithsonian castle, and the enlarged U.S. Capitol Building with its massive cast-iron dome. Another landmark of Corps engineering was the reservoir and aqueduct system that piped water into the district. Begun in 1853 and completed during the war, the Washington Aqueduct was a monument to its principal builder, Capt. Montgomery C. Meigs. The great aqueduct was also a highly politicized project that drove a wedge through the War Department and dramatized the willful persistence of an elite government bureau.

The trouble began with a collision of egos. Captain Meigs, a tall, sharp-tempered West Pointer from a powerful Philadelphia family, was a builder with little patience for pandering politicians; yet Buchanan's first secretary of war, John B. Floyd, was just such a man. Both men used family and friends to influence the pliant Buchanan. As the son of a famous physician and the nephew of a former cabinet official—John Forsyth, a prominent Jacksonian—Meigs represented the moderate lower North tier of the Democratic party, Buchanan's political base. Secretary Floyd, a former governor of Virginia, spoke for the bellicose South. Meigs was rigidly honest. Floyd was not above trading favors to advance a partisan cause. When he found public money to reward personal friends, Meigs protested to Congress. When Floyd attempted to purge the aqueduct project by firing non-Democrats, Meigs, defiant, held fast to a higher purpose: "Let our aqueduct be worthy of the nation, . . . Let us show that the rulers chosen by the people are not less careful of the safety, health, and beauty of the capital than the [Roman]

Washington on the eve of the Civil War,
showing the Corps' half-finished Capitol
dome project, broad avenues graded and
improved by the topographical engineers,
the army-planned and -supervised
Smithsonian Building (fire-proofed by
the Corps in 1853) and, faintly in the far
upper right, the army's Aqueduct Bridge.
Courtesy of the Office of History, U.S.
Army Corps of Engineers.

emperors who, after enslaving their nation, by their great works conferred
benefits upon their city." [87] An ambitious $2.3 million project and one of the
great hydraulic ventures of antebellum times, the aqueduct would tap the
Potomac at Great Falls, Maryland, and carry water into the district across
two magnificent spans. Meigs lobbied so hard for the costly project that
grateful city fathers gave the captain a silver kettle. Resisting compromise,
Meigs insisted on total control.[88]

   As Meigs lobbied for funds and battled the patronage system, he also
balanced the aqueduct project against his work on the Capitol Building,

another prestigious assignment. During the 1850s the captain supervised the Capitol's House and Senate additions. He also vastly improved the heating and ventilating system, hired European craftsmen and an Italian fresco artist, and designed the ribbing and cast-iron supports for one of the world's most spectacular domes. Although many applauded these Corps contributions, the tight supervision riled chief architect Thomas U. Walter, and soon the project was caught in a feud over who owned the architectural drawings and who deserved credit for what. Floyd used the dispute to discredit Meigs, but when Secretary Floyd demanded copies of the drawings in question, the captain flatly refused. Floyd then accused Meigs of insubordination. On November 2, 1859, an order from the War Department removed Meigs from the Capitol Building and confined the engineer to the aqueduct project alone.[89]

The myth of the engineer as an apolitical problem-solver remains basic to the scientific professionalism that still drives the Corps. That concept, however, overlooks the powers of implementation that allowed a builder like Meigs to tenaciously fight for his job. At one point the aqueduct builder stormed into the president's office during a reception hour. Buchanan did what he could. Unwilling to offend the Meigs family yet tightly bound to the Virginia wing of the party, Buchanan sided with Floyd, and eventually the president was forced to reproach the engineer for "evad[ing] the salutary checks which military discipline has interposed between . . . a subaltern and the Commander in Chief."[90] But Meigs still had strong allies in Congress. Working through Senators Jefferson Davis and Robert Toombs, a powerful man on the Finance Committee, Meigs had a short proviso added to the 1860 aqueduct appropriation bill. It mentioned Meigs by name and stipulated that the man in charge of the aqueduct be an officer of the Corps "not below the rank of a Captain."[91] The tactic was bold but ill advised. An affront to executive privilege, it alienated Buchanan, and Floyd was now free to punish the upstart captain. On July 24 a directive from Floyd brought in another builder and demoted Meigs to the job of inspector and aqueduct disbursement clerk.

Meigs fought on through the summer of 1860, but subsequent appeals to the White House were met with a cold rebuke. "The administrative supervision of the work on the Washington Aqueduct was legally and constitutionally confined to the War Department," said Attorney General Jeremiah S. Black.

"Will Congress pardon me if these [Floyd's] orders are enforced?" Meigs asked the Justice Department.

"He [Meigs] is taking too many branches of the Government under his care at the same time," was Black's dry response.[92]

Construction of the world's longest
masonry arch at the Cabin John
(aqueduct) Bridge, 1859. Courtesy of
the Office of History, U.S. Army Corps
of Engineers.

In the end the secretary demanded a court-martial, others called for
Floyd's resignation, and Buchanan was forced to decide. Banishment seemed
the lesser of evils. On October 11, 1860, an order from Corps headquarters
sent Meigs to Fort Jefferson on the Dry Tortugas, a desolate island outpost
about eighty miles off the Florida coast.

Andrew Jackson had warned that bureaucrats with long tenure in office
would undermine elected government by outlasting the politicians. Such
was the case with Meigs who returned to a hero's welcome in less than four
months. Floyd had already resigned, and the new secretary of war, a Meigs
supporter, reassigned the banished captain to the unfinished aqueduct project. Meigs also reclaimed his post as the Capitol Building's chief engineer. "I
have come back, Mr. President, to pay my respects," said Meigs to Buchanan
at a Washington party.[93] The president could barely respond. Three months
later Buchanan was in Pennsylvania, Floyd was behind enemy lines, and the
builder who stood up to corruption was the quartermaster general of the
Union army, a vital member of Lincoln's command.

Rock Creek (aqueduct) Bridge under construction, 1860, showing the curved cast-iron pipes. Encased in concrete, the pipes are now part of the Pennsylvania Avenue overpass between Georgetown and Foggy Bottom. Courtesy of the Library of Congress.

Below:
Cabin John Bridge, 1864. Courtesy of the Office of History, U.S. Army Corps of Engineers.

By then the aqueduct system was nearly complete. One of the largest waterworks of its kind—more advanced than the conduit system that supplied the city of Paris, and more than twice the capacity of Croton Aqueduct in New York—the Washington Aqueduct was a network of bridges and tunnels from Great Falls to Capitol Hill, a distance of sixteen miles. The most impressive piece of the project was the 450-foot aqueduct bridge across the Cabin John Valley in Montgomery County, Maryland. Designed by Meigs and his assistant Alfred L. Rives, an American graduate of the Ecole des ponts et chaussées, the main arch of the Cabin John Bridge, a 220-foot elliptical span, was the longest masonry arch in the world. Another amazing bridge crossed over Rock Creek on two 48-inch cast iron pipes. Although many hands made large contributions, most of the credit went to the Corps. General Meigs, triumphant and proud, ordered his name engraved on almost every structure, even the construction derricks.[94]

Meigs's aqueduct was the high note that capped an otherwise frustrating era. From Taylor to Lincoln—years of frontier surveying and westward expansion, an era of mixed success for army water construction—the battle for a unified "system" of magnificent projects had been fought and ultimately lost. River snagging had virtually died under Pierce and Buchanan. The search for a rail line to the Pacific had derailed in a spiteful debate over a route through the southwestern desert. And when civil engineers campaigned to reform the army contracting system, Totten, ducking the scandal, escaped on a tour of Europe. West Point, meanwhile, was losing its long-standing preeminence as the Harvard of construction science. By 1866 at least sixteen civilian schools were offering engineering or applied science degrees.[95]

Still the West Pointers had won smaller battles. Playing the War Department against powerful committees in Congress, the engineers had learned to survive in the cracks between jurisdictions. As planners for Congress, the engineers could shape and sometimes stonewall harsh directives from a Floyd or a Davis, yet the same builders were the president's men, agents of the executive branch. In this way the era of great expectations was the seed of a vast organization. Nearly a hundred years after Meigs stood up to Buchanan, the Corps, said a critic, had become "a veritable Rock of Gibraltar," a stubborn child of its own traditions, an agency that used engineering and science to reward the friends of the army and cultivate local support.[96]

3. The Imperial Corps

The era of Meigs also revealed a face of the scientific profession that Congress tried to ignore. Clannish and controlling, an imperial corps d'elite, the engineers were not quite the despotic monster that Alden Partridge envisioned, but wealth and family connections did open doors for polished West Pointers. The image of the army builder as a privileged bureaucrat, although a caricature, was based on a partial truth.

Family, most of all, was the path to the free education that bought membership into the club. Among the top engineer clans—the Swifts, Mansfields, Mahans, Aberts, McClellans, Lees—a West Point degree was almost a birthright, a tradition passed from fathers to sons. Meigs, for example, won a West Point appointment with help from his father's sister, Clara Meigs Forsyth, the wife of Jackson's secretary of state. When the engineer tried to pull similar strings for his bright eldest son, John Rodgers Meigs, Floyd blocked the appointment. The father then appealed to Buchanan and strengthened his son's application with letters from powerful men. One letter from Jefferson Davis included bipartisan endorsements from Senators Stephen A. Douglas (Illinois), Zachariah Chandler (Michigan), William Bigler (Pennsylvania), Albert G. Brown (Mississippi), John Slidell (Louisiana), Alfred Iverson (Georgia), William M. Gwin (California), and others. Floyd relented.[97]

Some cadets with the proper connections were channeled toward engineering whether they liked it or not. As the son, nephew, and cousin of army topographers, West Pointer James Abbott McNeill Whistler, the future artist, seemed predestined for an army career. Still Whistler, a reluctant scholar, had no patience for science. Once a chemistry professor asked the cadet about the properties of silicon. "Silicon, sir, is a gas," Whistler responded. "Had silicon been a gas," he later remarked, "I would now be a major general."[98]

Whistler's sarcasm indicated the value West Pointers placed on science, and it poked fun at the grand expectations of a rising professional class. Class and status, nevertheless, were taboo subjects in official reports. When Senator Davis organized a study of the five-year course at West Point, the massive investigation dodged the nepotism/patronage issue. Printed by Congress in 1860, the report mentioned "the poor" only once, and then just to praise "the eminently popular, republican, and equal conditions of admissions which open the military career to the ambition and genius of all the sons of the Republic."[99] Elsewhere the Davis investigation noted Spartan living conditions. The mess food was bad. The barracks were cold. Therefore, according to West Point logic, the school was democratic.

Still, any outsider could see that class and family connections helped bond the inbred profession. "Never marry," said Robert E. Lee, "unless you can do so into a family which will enable your children to feel proud of both sides of the house."[100] This sort of advice made sense to men like Totten, civil engineer Thomas A. Davies, and their engineer cousin Joseph K. F. Mansfield—all tied through schooling and marriage. A Graham married a Meade. Humphreys married his cousin. A daughter of M. C. Meigs married a nephew of Zach Taylor. Mary Roberdeau Swift, the first wife of topographer George Whistler, was also the niece of both Isaac Roberdeau and Joseph Swift. Whistler's second wife, Anna Mathilda, was the sister of fellow West Pointer William McNeill. The same Anna was the famous matron in black of the "Whistler's Mother" portrait that now hangs in the Louvre.[101]

Engineer wives became, as one scholar put it, "a sexual resource" that empowered army construction by linking builders to prominent men.[102] Thus engineers Sanders and Frémont both married congressmen's daughters. Engineer John Newton, a congressman's son, married Anna M. Starr, the daughter of Jonathan Starr, a leading Connecticut banker. Canfield's wife, Mary Cass, was the daughter of Michigan governor Lewis Cass. Emory married Matilda Wilkins Bache, the daughter of A. D. Bache. Meade joined Emory's topographical corps through the influence of his Whig brother-in-law Henry A. Wise, a congressman from Virginia. McClellan married his commander's daughter, Mary Ellen Marcy. Their daughter courted the grandson of John Quincy Adams. And Meigs, to cite a final example, married Louisa Rodgers, the daughter of the Comdr. John Rodgers and the sister-in-law of topographer John N. Macomb. Meigs was not above using his wife to win a reprieve from Buchanan. When Meigs was banished to Fort Jefferson, he left Louisa behind as a "silent" yet "living protest against a good citizen's exile." [103]

Given the importance of family and marriage, it stands to reason these literate builders would have written about the mothers and wives who bound the tight-knit profession. Yet the published record on women's affairs was like the martyred protest of Louisa Meigs—nearly silent.

Even when socialite Harriet Balch Macomb was touched by a Washington scandal, the incident was quickly dismissed as a frivolous woman's concern. Mrs. Macomb, the chief engineer's second wife, was one of many who refused to acknowledge Peggy O'Neale Eaton, a former barmaid and the new wife of John H. Eaton, Jackson's secretary of war. The Eaton affair enraged President Jackson. Yet cool-headed General Macomb, one politician recalled, "took such things lightly," more amused than annoyed.[104]

Socialite, sexual resource, object of devotion—an engineer's mother or wife could also serve the profession as a publicist or confidante. Perhaps the most famous example was Jessie Benton Frémont, the Missouri senator's daughter and the topographer's wife. A gifted travel writer, Jessie embellished her husband's 1844 Oregon Trail report, one of the most widely read books of its time.[105]

Another woman briefly in the public spotlight was Catlyna Pearson Totten, a genial New Yorker. Since her husband, the chief engineer, was removed from most of the action during the Mexican War, Catlyna kept Corps headquarters informed by passing around a battle report in a personal letter from one of Scott's excellent staff officers, Robert E. Lee. Lee's letter to Catlyna Totten embarrassed the Corps when it appeared in *The Washington*

Union. More often than not, however, the mothers and wives of the male profession were confined to symbolic roles. Some were like the goddess statue called *Armed Freedom* that Meigs placed on the Capitol dome: a feminine symbol of civilization, a metaphor for progress through militarized science. At Hell Gate, for example, where Maj. Gen. John Newton packed fifty thousand pounds of explosives into a dangerous rock, the engineer invited women and enlisted his own daughter to dramatize the event. Two-year-old Mary Newton touched the electric button that set off the monstrous blast.[106]

Civil engineering also was a male, elitist, and somewhat inbred profession whose members increasingly tried to exclude the mechanics who rose through the rank.[107] Seldom, however, was civil engineering hit with the populist anger that vilified the men of West Point. Editorials in *Engineering News* and *Iron Age* censured the Corps engineers as an "egotistical" and "pseudo-scientific class."[108] One Gilded Age critic returned to the bureaucrat-despot theme first developed by Partridge: "river and harbor works [had become] an inheritance belonging to their [the army's] self-constituted military aristocracy—an aristocracy quite ungenerous and overbearing and having no right to its pretensions."[109] Civilian attacks on the Corps were again a righteous crusade.

The Great Rebellion only postponed the inevitable confrontation between rival professional camps. A time of staggering growth for the regular army, the Civil War was indeed an "organizational revolution"—a bureaucratic war, a turning point in the campaign to consolidate government and federalize public works. Fort builders were overwhelmed. In a war of long range rifled artillery and iron ships that could storm into fortified harbors, the Corps wasted resources on the "third-system" coastal defenses, the fort network planned by Bernard. In a booming era of civilian construction, a time when the U.S. census recorded a ninefold increase in the civil-engineering profession, 1850 to 1870, high-ranking West Pointers still hoped to exclude non-West Point engineers.[110]

Age was part of the story. Colonel Abert, near death at age seventy-three, was chair-bound in his Washington office and "half-dazed," according to one engineer.[111] Forced out in September 1861, the senior topographer was briefly succeeded by another great name from a previous era—Stephen Long. Meanwhile the ranking fort engineer was "conservative in all his views" and, Meigs reported, "old in subordination."[112] Chief Engineer Totten, a month older than Abert, sapped wartime mobilization by snubbing civilians and clinging to faint lines of distinctions between topographers and fort engineers. Congress brushed aside Totten's concerns with legislation that merged the two engineer organizations. Signed by Lincoln on March 3,

Canal at Delta Point, 1862. Laboring in blistering heat, 1,200 slaves toiled alongside Union soldiers in an aborted attempt to bypass Confederate guns by cutting through a bend in the Mississippi River near Vicksburgh. Courtesy of the Boise Public Library.

1863, the act abolished the twenty-five-year-old topographical corps. Officially, the new Corps of Engineers was stronger than ever before with a brigadier general, 105 engineer-officers, and four companies of engineer-soldiers—855 men in all. But Lee and fourteen others had defected to the Confederate army. With losses and transfers, the number of commissioned engineers in the regular army was actually fewer than before the war.[113]

When Totten died in 1864, Brig. Gen. Richard Delafield, the new chief, made a bad situation worse. At age sixty-six the former West Point superintendent had served fifty years in the army without ever experiencing combat. Widely disliked, Delafield cut budgets, spied on his men, undermined a successful photography and mapmaking program, and otherwise squelched innovation. "If Totten was a passive force," said historian Philip L. Shiman, an expert on Sherman's march, "Delafield was regressive."[114] One engineer called the penurious chief "a worthless unreliable old fellow."[115] Another hoped Delafield would contract diarrhea and die.

Still, the Civil War did inspire some remarkable engineering: a photographic process for reproducing maps, a two-mile steamer canal through a bend in the Mississippi, pontoon bridges, timber highways, fortified railroads, strategic innovations with blockhouses and entrenchments—all projects born in the field despite bureaucratic interference. Some of the best engineering work was done by civil "volunteer" engineers and thousands of common builders. Volunteers Haupt and Ellet both worked apart from the regular army; yet Haupt launched the highly efficient railroad construction corps, and Ellet designed an excellent fleet of ironclad battle rams. Combat engineers also made large contributions. At the Battle of Gettysburg the break for the Union came when Maj. Gen. Gouverneur Warren, Meade's chief engineer, saw the value of Little Round Top, a critical hill. Infantry arrived just in time to turn back the rebel attack. Gettysburg—like Antietam, Manassas, Fredericksburg, and many of the bloodiest battles—was a brawl among determined West Pointers with engineers in command on both sides. By April 1865 twenty-five antebellum Corps and topographical officers had become major generals in the Union army. Of the eight highest-ranking West Pointers in the Confederate army, three were once engineers.[116]

In the end the four years of carnage changed everything and nothing at all. A shock to the two-party system, the war discredited, for a time, the Democratic state's-rights position. Yet the nation was still of two minds about the value of army construction. Beast and benefactor, the Corps, still an agent of centralization, remained a target of the populist rage against science, big government, and the army's imperial power.

The first civil-works task for the army was to deal with the damage of war. Fallen bridges, twisted rails, snags, sunken vessels, and sandbars were forestalling the promise of reconstruction, taxing commerce as never before. The Delta country needed levees. Baltimore needed dredging. The Rock Island and Des Moines rapids bottled up the Iowa grain trade. Philadelphians wanted a thirty-foot cut to the ocean. Lake shippers campaigned for lighthouses, port improvements, a Fox-Wisconsin canal, a St. Lawrence seaway, and a passage through the mud near Detroit. As the South joined the West and the North in the new age of national projects, Congress, a year after the war, appropriated $3.6 million for water improvements and surveys at forty-nine sites. Bills grew in size and expense until the office of the chief engineer was fielding about five hundred assignments. In 1882 the annual expenditure for rivers and harbors topped $18 million. "Let us scatter this money out," an Indiana politician advised.[117]

The free-spending era of pork boosted army construction but left the Corps on the defensive against threats to its traditional turf. One blast at the science tradition was the 1866 law that detached West Point from the Corps. Another challenge to the engineer bureau was the 1867 revival of a proudly civilian organization, the American Society of Civil Engineers. Government scientists also vied for Corps jurisdiction. In 1875 the Smithsonian's Ferdinand Hayden vowed to "crush" the Corps reconnaissance program, and

John Wesley Powell, the Grand Canyon explorer, said the army's meandering methods were superficial and obsolete.[118]

Perhaps the fiercest attack in this new era of professional combat was the challenge to Corps jurisdiction from the civil-works contractors themselves. An ancient conflict, the tension between soldiers and toiling civilians was aggravated by the postwar construction boom. Suddenly there was more funding for rivers and harbors than prudent West Pointers could spend. In 1873, for example, Maj. William P. Craighill of the Baltimore district was embarrassed to receive $5,000 more than requested. Again in 1882, of the 371 improvements approved over a presidential veto, 18 projects were funded despite unfavorable reports from the Corps. Another 16 projects had federal money before the sites were even surveyed. The Corps managed the deluge through a river and harbor division with dozens of special bureaus: harbor boards, lighthouse boards, an irrigation commission, a waterpower commission, a Nicaragua canal survey—all headed by soldiers. Hundreds of civilian "assistant engineers" now worked on government contract. Hired and fired at the whim of the War Department, they were "bread and butter men," an expendable work force of practical builders, a subordinate caste. "The government engineers are divided into two distinct classes," a magazine writer complained. "Never in our civil service was the line between politician and worker half so rigidly drawn."[119]

Civilians despised everything about the system. They resented the web of red tape that drove up the cost of construction, the junior officers who barked orders at experienced builders, the fair-weather superintendents who took credit for water construction without assuming the financial risk. Moreover, the critics insisted, West Pointers were poorly trained. At a time when engineer schools at Yale, Dartmouth, and Rensselaer were developing three- and four-year courses, postwar West Point offered less than two years of pure engineering. Some army textbooks were forty years old.[120]

Yet the army caste system survived. It persisted because of the Corps' strong ties to Congress and the fame of a few West Pointers who were still the nation's recognized leaders in canal engineering and river and harbor design. Submarine explosives, iron-hull snag boats, suction-type dredge boats, a pneumatic process for building underwater in an airtight box—all were Corps innovations. Postwar army builders still had a flair for colossal construction that was functional yet monumental. From 1869 to 1872 a decorated major of engineers, Godfrey Weitzel, enlarged the Louisville and Portland Canal with the world's widest locks. Ten years later Weitzel broke records again with the 515-foot lock at Sault Ste. Marie. Another lock and dam pioneer was Col. William E. Merrill, the father of Ohio River canaliza-

Davis Island Lock and Dam, 1885.
Courtesy of the Office of History, U.S.
Army Corps of Engineers.

tion. An avid student of French and German hydraulics, Merrill wanted to pool the Ohio with movable dams that closed in order to back up the river and opened to flush traffic downstream. His prototype was the folding "wicket" dam below Pittsburgh at Davis Island. Completed in 1885 under project engineer Frederick A. Mahan, Merrill's able assistant, the Davis Island Lock and Dam was a 1,223-foot wall across the Ohio with a masonry pool for steamboats and barges—600 feet long and 110 feet wide, the largest lock chamber in the world.[121]

Like many of the Corps' most spectacular projects the movable dam at Davis Island was an adaptation of a Parisian design. European technology also endured in the Mississippi levee system. Since 1841, when Capt. Alexander Swift saw French and Dutch workers laying straw mats on fragile sand dunes at Pont de Grave, the army had taken an interest in woven or matted revetments that confined a flood-prone river by protecting its crumbling banks. Engineer Charles A. Hartley, the British father of flood control on the Danube River, further popularized bank reinforcements. A. A. Humphreys believed protected embankments could contain the lower Mississippi if hundreds of miles of levees were raised to a uniform height. Revising his Delta

Building willow fascine mattresses along the Mississippi River, about 1885. Courtesy of the Office of History, U.S. Army Corps of Engineers.

Below:
The Mississippi's twisted path to the ocean, 1863, looking south from New Orleans. Courtesy of the Historic New Orleans Collection.

survey in 1866, Humphreys reported the ruin of the antebellum levee system: 107 miles of damaged levees in Louisiana alone. Congress appropriated $200,000 for levees in 1873, and soon the Corps was experimenting with European-style revetments—fascine levee mattresses called "brush rafts." Developed by Col. James H. Simpson at Sawyer Bend above St. Louis, the mats were woven with long willow poles, bound with wires or rope, and anchored underwater with stones. Maj. Charles R. Suter improved the willow design in tests on the lower Missouri in 1877.[122]

The levee program was also a chance for engineers to make a political point about the value of centralization. In 1874, a bad flood year, Congress organized a five-man levee commission, and soon the Corps was again calling for federal action in an exhaustive report. Compiled by Maj. Gen. G. K. Warren and published by the House in 1875, the investigation blamed local government for "fatally defective" levee organizations, and it proposed six regional districts with chief engineers and boards to settle disputes.[123] Congress tabled the plan. Almost two years later, however, the levees resurfaced as an issue in the contested presidential election of Republican Rutherford B. Hayes. As House Democrats filibustered to block the inauguration of Hayes, the southern delegation defected, trading votes for the promise of aid. What the South wanted besides an end to Reconstruction were harbor repairs, a railroad to San Diego, and a well-funded levee program. On June 28, 1878, Congress and Hayes agreed on a Louisiana plan for a joint military and civilian Mississippi River Commission, a bureau to plan and construct levees.[124]

West Pointers cited the levees as proof that river construction still needed soldiers to promote standardization. Civilians countered with a long list of fiascos. In 1868, for example, a Corps "floating dam" at the mouth of the Susquehanna lasted only a week before a storm ripped into the structure. At the Washington Aqueduct, the engineers wasted $1.3 million on a short-lived tunnel project. Another $937,000 was lost in a failed attempt to deepen Florida's St. Johns River. In 1877 the long-awaited Des Moines Canal, a twelve-year $4 million project, opened for three weeks, closed for repairs, opened for two weeks, then closed again. "The government [has] been twelve years building the [Des Moines] canal and had given the people nothing," said W. H. Russell, a spokesman for the St. Louis Merchant's Exchange.[125] Canal contractors went on hunting expeditions while the grain rotted on the docks.[126]

The mud-filled mouth of the great Mississippi was perhaps the biggest frustration of all. With ships grounded on bars and only fourteen to eighteen feet of water over the deepest pass through the Delta, New Orleans was

James B. Eads, engineer and entrepreneur. Courtesy of the Missouri Historical Society.

virtually closed to deep-water freighters. In 1866 a British cotton ship with a sixteen-foot draft was stuck in the mud for nine days. Dredging seemed the cheapest solution. With $223,000 from Congress in 1867, the Corps worked with Boston shipbuilders to develop a three-engine boat with mud scrapers and two fourteen-foot screw-blade propellers, one at each end. But the dredge was no match for the mud bars. A pump cracked. Blades broke off the propellers. And after ten disappointing months in a pass below New Orleans, the dredge had made hardly a dent. *Essayons,* the name the Corps gave to the boat, made a mockery of the agency's motto: "Let us try." [127]

The man most tarred by the Delta problem and other frustrations was the Corps' reigning expert on river hydraulics—Brig. Gen. A. A. Humphreys, the postwar chief engineer. A theoretician more than a builder, Humphreys, unlike Delafield, was a battle-wise engineer who brought out the Corps' most bellicose traits. The man relished combat. Ordered into a deadly ravine at the Battle of Fredericksburg, Humphreys, a division commander, had led an impossible charge at a line of Confederate rifles behind a fortified wall. Hundreds died in an instant. Mounting a wounded horse, Humphreys dashed through the field shouting orders, rallying troops for a second attack. "He was very pleasant to deal with, unless you were fighting against him, and then he was not so pleasant," a battlefield colleague recalled.[128] From

Eads Bridge at St. Louis, 1876. Siding with the steamboat and ferry interests, Chief Engineer A. A. Humphreys and a five-man board of high-ranking West Pointers, all Corps engineers, vainly tried to halt construction, denouncing the steel arch bridge as a hazard to navigation. The structure, nevertheless, has been hailed as "the most remarkable engineering job in bridge history." Courtesy of the Missouri Historical Society.

1866 to 1879, as the Corps' eighth chief engineer, Humphreys fought losing battles to iron-plate existing masonry fortifications and to dominate the western surveys. In 1878 he also took on the National Academy of Sciences. When civilian scientists endorsed a plan for a new geological bureau, Humphreys, although a charter member of the science academy, resigned from the organization. The protest touched off a pointless debate over which arm of the government could map the most miles per dollar.[129]

Throughout these battles for Corps jurisdiction the eyes of Congress were locked on the troublesome Delta. The dredging program had stalled, a Minnesota senator went south to investigate the high cost of dragging boats off the mud bars, and a visionary jetty builder, James B. Eads, took on the Corps and its science by defying the chief engineer. Eads was everything Humphreys was not. Banker, venture capitalist, a self-taught river expert, and a hero to the Delta people, Eads was among the last of the great builder-inventors without any kind of formal degree. His problem with the Corps began in 1873 when Humphreys convened a board to study the Eads Bridge, a huge and innovative three-arch span across the Mississippi at St. Louis, the first railroad bridge framed in steel. Six years into the project the Corps said the bridge was too low for steamboats. The structure would have to come down. Eads rallied support from St. Louis, appealed to Congress and President Grant, and finished the bridge as planned in 1874. The builder then struck back at the Corps with an astonishing proposition. He would solve the ancient problem of the Mississippi mud bars with jetties of his own design. Walling a shallow pass in the Delta, he would narrow the Mississippi, increase its current velocity, and force mud out to sea. If the jetties cut a twenty-foot depth within thirty months, Eads would receive $500,000. The deeper the channel, the more money he was to receive—up to $5.2 million. If the project failed, however, Congress would pay nothing at all.[130]

Humphreys staked his high reputation on the promise that the jetties would fail. Citing his own great tome on the river, the 1861 Delta survey, the chief engineer made two principal points. First, said Humphreys, the "lumps" that blocked navigation were a hard blue marblelike clay more dense than the sides of the passes. Speeding the current with jetties would only cut soil from the banks. Second, the scouring jetties of Europe all depended on "littoral" currents that swept silt out to sea. "No such current has been discovered at the mouth of the Mississippi," said Capt. Charles W. Howell, the Corps' man in the Delta. "Jetties would have to be built further and further out, not annually, but steadily every day of each year, to keep pace with the advance of the river deposit into the Gulf."[131]

What the Corps wanted instead of the jetties was a $7.4 million canal. In 1874 a seven-man board of Corps engineers (inspired, perhaps, by Britain's recent success at Suez) endorsed Captain Howell's plan to bypass the Delta with a deep canal from the river eastward to Breton Sound. The one dissenting voice on the board was its senior member, Col. John G. Barnard. An old hand in the Delta who, years before, had published a study of jetties, Barnard preferred a natural path to the ocean. This lone note of caution and Eads's sincere presentation left just enough doubt in the House to delay a vote on the canal. After a second board of inquiry and a new round of numbing debate, Congress, on March 3, 1875, went with the jetties. The one big disappointment for Eads, though, was that Congress, following the recommendation of Barnard and visiting engineer Charles Hartley, had insisted on the South Pass of the Delta. Eads had wanted the bigger Southwest Pass. By mid-June, nevertheless, the first jetty crew was sinking a line of timber through the reeds and swarming mosquitoes of shallow South Pass.[132]

Humphreys resisted the work to the end. Attacking through his subordinates—Howell, John Newton, and others—the chief orchestrated a campaign to smear Eads and Barnard. The low point came in the spring of 1876 as Eads struggled to refinance his South Pass Jetty Company, a stock corporation. As

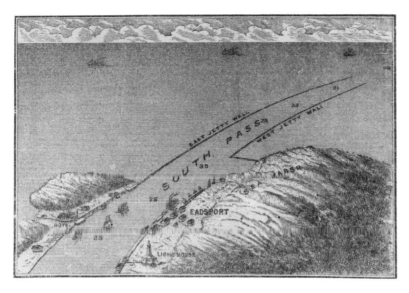

Above:
The mouth of the Mississippi Delta, showing the Eads Jetties at South Pass, around 1890. Courtesy of the Historic New Orleans Collection.

Bird's-eye view of the Eads Jetties, around 1890. Courtesy of the Historic New Orleans Collection.

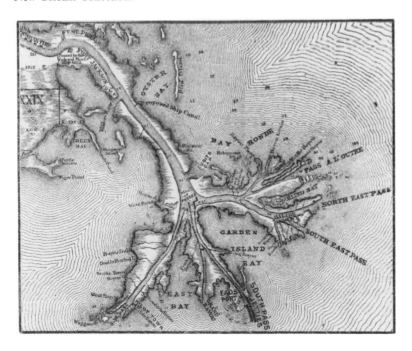

(a) Cross-section of the South Pass, showing channel depths before and after the jetties, 1884; (b) seaward end of the Eads Jetties, 1884. Courtesy of the Historic New Orleans Collection.

potential investors toured the half-finished project, H. C. Collins, one of Howell's dredge boat assistants, leaked a dubious chart to the press. The chart purported to show that a new shallow bar was forming at the head of the jetties. Eads called it a lie but the damage was done. Investors panicked, stock fell, and a fickle New Orleans press hinted that Eads was a fraud. "It appears," said Walter M. Lowrey, an early historian of the project, "that the Chief of Engineers and Howell were determined to ruin Eads even if they had to lie to do it."[133] The more Humphreys attacked, however, the more Eads supporters in Congress dug in their heels. Praising Eads as "this benefactor of the Mississippi Valley, of the United States, and of the World," Congressman Omar D. Conger of Michigan, a Republican spokesman for river improvements, convinced the House to keep the project afloat with a $500,000 advance in inflated government bonds.[134]

By 1879 it was clear that Eads would soon finish the jetties even if their effect on the current and mud lumps was still an open debate. The project was already a tourist attraction. An ingenious design, the South Pass jetties were a walled shipping channel about one thousand feet wide and more than two and a quarter miles long. Its walls were willow rafts prefabricated in the marshes, floated downriver, stacked underwater, and compressed into a hard dike with concrete blocks. Even before the 260-ton blocks were hoisted onto the structure, the powerful Mississippi, energized by the jetties, was excavating a navigable trench. On July 10, 1879, an engineer verified a depth of twenty-six feet at the head of South Pass and no less than thirty-one feet throughout. The project, barely, had met the terms of the contract. Eads had gambled and won.[135]

"The work is done," said the *Daily Tribune,* reporting the triumph from New Orleans. "There is no achievement of mechanical genius which compares with it in the splendor of its economies or in the magnitude of its results."[136] Eads immediately began promoting an even more magnificent project—a transcontinental ship-railway across southern Mexico. Humphreys retired.

To the extent that builders regarded the jetties as a professional confrontation—a duel between Humphreys and Eads, between army science and civilian technique—the episode painted West Pointers as small-minded men of routine. But the Corps saw another side of the conflict. In 1886, when the American Society of Civil Engineers published a forum on the jetty affair, Colonel Merrill claimed that the South Pass shipping channel was smaller than promised: it was too shallow in places, and the channel had never reached the required width of three hundred feet. Others characterized Eads as a pretender and slick entrepreneur who circumvented the contracting system by going directly to Congress. "A legislative body [Congress] is notoriously a bad organization for making contracts," Brig. Gen. Cyrus B. Comstock explained.[137]

In Louisiana, meanwhile, the booming trade spoke for itself. New Orleans was again a hub of international commerce. Previously the eleventh busiest port in the nation, it grew bigger than all but New York. Galveston now wanted jetties; so did Charleston, Savannah, Oakland, and about thirty sites in New England. The Corps, too, was a convert. Ten years after Eads sold the idea of jetties to Congress, an engineer captain explained that the concept of a South Pass shipping channel was a Corps innovation that dated back to Fillmore's time. And in 1924 a chief engineer maintained that Eads had "merely carried out plans which had been previously discussed."[138] It seemed as if the jetty path to the ocean had been the Corps' idea all along.[139]

4. Siege at the Gates of Verdun

A symbol of strength and tradition, the castle is a fitting insignia for America's largest and most entrenched engineering organization, an agency often besieged. First used on Corps epaulets and belt plates during the 1840s, the castle design may represent a fortified city in Europe, possibly the gates of Verdun. Agency historians say the turreted castle symbolizes valor in combat.[140] Like many symbols, however, the castle invokes an image beyond its avowed meaning: an image of power and militarism, a tribute to the lost nobility of the fighting savant. Today the red castle flags at thousands of Corps installations are the nostalgia of an agency searching backward for mission and purpose. Rooted in martial tradition and responding to cues from within, the Corps, its flags and insignias remind us, is a product of an inner culture much older than our political system and sometimes outside its control.

Castles, in medieval times, were also symbols of the landed gentry, and thus the insignia of the Corps takes us back to attacks on the "privileged order" that, the antebellum critics maintained, regimented the Republic, fostering a despotism that corrupted American life. Shadows of that evil persona still haunt the civil-works organization. New Dealer Harold L. Ickes, writing in 1951, characterized the Corps as "the political elite of the army," a "ruling class."[141] To Gene Marine, author of the 1969 best-seller *America the Raped,* the agency was "the most nearly untouchable empire in the United States, as powerful in its field as the FBI or the CIA and as difficult to oppose."[142]

The Corps, of course, has always had its defenders—writers like political scientists Jeanne Nienaber Clarke and Daniel McCool who praised the agency for its flexibility, calling it "sincere, swift, and impressive."[143] A 1979 Brookings Institution study singled out the dam-building Corps as a "possible exception" to the rule that large bureaucracies are like slow-moving glaciers that crush innovation.[144] Hardly despotic, the Corps, say its defenders, is the epitome of responsiveness through decentralization—a grassroots organization that caters to community needs through its eleven U.S. division and thirty-seven regional districts. Hardly militaristic, the civil-works side of the Corps answers to the army's assistant secretary for civil works, a civilian. Today the engineer-officers are a small minority of about 900 in a civilian sea of about 26,000 civil-works employees—although the divisions and districts are still led mostly by West Pointers. In 1992, for the first time since the Truman administration, the chief of the Corps, a lieutenant general, was a non–West Point engineer.

The Corps, therefore, is civilian in many respects; and yet, if there is a point on which the agency's defenders and detractors can mostly agree, it is that the Corps' leadership is a product of a professional culture rooted in a military past. But how? Maj. Gen. Harry Taylor, a 1920s chief engineer, said the agency's debt to the army was honesty, integrity, and "complete freedom from political influence."[145] Others say combat experience gave the can-do organization a spirit of activism. One patriotic account calls the Corps a "fighting elite," citing a legacy of "resolution, fortitude, and vision, . . . not

Left: Corps castle design, 1840, possibly patterned after the city gates of Verdun, France; *Above:* the castle logo, 1980s. Courtesy of the Office of History, U.S. Army Corps of Engineers.

only a tradition of success in military engineering, but a discipline of growth, progress, and scientific evolution that was part of the expanding strength of America."[146] In 1966 a Senate public-works committee reported that "military training" and the "distinct military character" of the civil works organization gave the Corps its management style.[147]

But did that management style shape the character of the nation? Did it seed factory capitalism? Did it "spill over," as the economists say, into the industries and institutions that govern our late-twentieth-century lives? Americans disagree. Many historians are convinced that the men of West Point unleashed a meticulous yet practical science that accelerated America's growth as an industrial power. Railroads, engineering education, scientific mapping, iron-framing, photographic printing, and the "American system" of factory machine-tool production have all been cited as West Point spillovers into civilian life. Writing about the rise of railroads and Carnegie's steel empire, Harold C. Livesay claimed that West Pointers like Isaac R. Trimble and George B. McClellan were the first American masters of elaborate and highly specialized bureaucratic organizations. Historians of the factory system have found military thinking behind scientific management and assembly-line regimentation.[148]

Others say the economic effects of military spillovers and spin-offs are easy to overstate. Before the twentieth century, wrote Joel Mokyr in *The Lever of Riches* (1990), "the needs of military technology seem to have intersected little with those of civilian production."[149] And recently Martin van Creveld has argued that "military organizations have tended to be even less flexible than most large bureaucratic structures."[150] Subordination, secrecy, discipline, and a rigid social structure are military characteristics that, van Creveld maintained, have stifled civilian invention and restricted the free flow of ideas.

The history of rivers and harbors is rich and varied enough to support both positions. The Corps, in the modern literature on water construction, remains Jekyll and Hyde: the catalyst of capitalism and a threat to the capitalist system, a bold innovator and a brake on the wheels of change.

Unquestionably the waterways-construction mission promoted growth and industrialization. Dikes, breakwaters, piers, canals, and other army achievements have dammed and diverted about twenty-six thousand miles of navigable waterways, turning North America into one of the world's most extensive hydrological systems. By the time of the Civil War about 90 percent of cotton bought and sold in New Orleans was shipped to market by water. By 1900, after the Corps opened long stretches of the Mississippi-Missouri system, most of the good farmland on the Great Plains was within forty miles from a navigable channel. No project symbolized a century of West Point achievement better than Panama's monumental canal. Reorganized and finished ahead of schedule by West Pointer George W. Goethals, a Corps lieutenant colonel, the canal was held up as proof that the army engineers had skill and perseverance enough for any technological challenge.[151]

Still, the Corps seemed hesitant to embrace too much innovation, and thus a good argument can be made for William Ashworth's picture of water agencies as "hidebound, unimaginative, and smothered so thoroughly in political horsefeathers that it becomes a thing of wonderment they can move at all."[152] Those who think the Corps fits that description often point to the aftermath of the feud between Humphreys and Eads. Challenged by a swelling civil-engineering profession—and also by new scientific agencies such as the Mississippi River Commission and the U. S. Geological Survey, the Corps retreated to the position that "levees only" could protect the lowlands against inundation. Historian Martin Reuss has maintained that the Corps resisted flood dams, flood outlets, and river cut-offs because, back in 1861, topographers Humphreys and Abbot had vetoed these innovations in their famous report. Not until 1917 did the agency accept a responsibility for protecting the flood-torn lower basins of the Sacramento and Mississippi, and not until 1936 did Congress extend that protection to the forty-eight states. Even then the Corps leadership questioned the value of dams.[153]

Perhaps the most that can fairly be said about the Corps' management style is that it could be proactive or reactive depending on what was at stake. Forced into uneasy cooperation with the civilian engineering profession, the agency quickly embraced some civil-works innovations—snag boats, dredge boats, iron-framing, steel foundations—but the same organization fought hard against technologies and reforms that seem to compromise army planning and challenge Corps expertise. The engineers, nevertheless, remained forward-looking enough to guide the nation through an important transition. They helped Americans fear science and bureaucracy less than partisan gridlock. In the process they taught the Republic to value rivers and harbors as technological systems—as thousand-mile levee networks, as regulated highways of commerce that spread prosperity to the heartland, unifying the states.

Effects of that comprehensive approach ripple through our federal system. Corps emphasis on system and order still fuels the Whiggish conviction that state and local projects are chaotic and inconsequential. Corps ideas about scientific control also contribute to a bigger-is-better bias that fosters monumental construction but veils some of the cost. Only in the last generation have the engineers been forced to concede what their critics were beginning to see during the era of the general surveys: economic progress has unplanned consquences, and even the most scientific attempts to turn rivers into technological systems tip a delicate balance between rushing water and earth. Thus piers that block floating ice silt with shallow deposits. Reservoirs cool the water and dramatically alter the food chain. Levees straighten and accelerate rivers, increasing erosion downstream. Dams rob the ocean of nutrients, block migrating salmon, and threaten important fisheries from Oregon to Chesapeake Bay. Today the Corps spends millions to mitigate the destruction unleashed by the great national projects of previous decades.

And so something about the Corps stays constant as the nation around it evolves. An aging child of West Point professionalism, the civil-works organization turns slowly in new directions without escaping the conflicted traditions that anchor engineers to their past.

Formative Conflicts

~~~~~~

Born in a lively debate over the nature of our federal system, the U.S. Army
Corps of Engineers emerged from the formative conflicts that shaped the
Republic at large. The first, a cultural conflict, pitted a French tradition against
British and frontier ideals. The second, a technological conflict, was the tension between scientific methods and the craftsmanship learned in the field.
The third, a political conflict, was a debate about the Republic itself. If, as
Hamilton said, the constitutional experiment was "necessarily a compromise
of dissimilar interests and inclinations," then the engineers extended that
compact.[1] Cheering federalism, preaching unity through science, industry,
and modernization, the Corps braided the patchwork of dissimilar interests
into a nationwide network of power.

Today, in our time of $2 billion canals and rivers that run backward, the
Corps still manages water yet the agency's origins seem strangely remote. Social scientists worry about "technocracy" and "the professional state" as if the
idea of the scientific expert in government was a twentieth-century concept.
Not until Gifford Pinchot and the conservation movement did the politics of
efficiency seem to reorder the executive branch. Revolution is often the word
for bureuacracy's swift transformation, as in "the managerial revolution," the
urban "public works revolution," or the Progressive era "revolt of the engineers" that led to the big-government "social revolution" of the 1930s.[2]

Doubtless the influence of bureaucracy has swelled in our lives, but the
rise of scientific professionalism was a gradual process. Statistical reporting,
cost-benefit equations, specialized field offices, standardized forms, and standardized regulations were antebellum tactics. Sold as the solution to partisan

gridlock, the professional state was a response to the chaos that stalled public works.

Nationalists are still fighting that chaos, and their scientific solutions still fire a heated debate. At one extreme are government's staunch defenders, those who refer to "the public good" and "the general welfare" as if program implementation were politically neutral, as if planning were impartial, as if bureaucracy was, as Samuel Florman maintained, "supremely sensitive to every wish of Congress." [3] At the other extreme are the decriers of science as a tool of technocratic consolidation. Both positions underestimate enduring confusion over whose science is the most impartial and whose experts should plan. The case of the Corps suggests that professional division, more often than not, has fueled the political combat that derailed national projects. And if bureaucracy has become, in Thorstein Veblen's words, "a somewhat fantastic brotherhood of over-specialized cranks," the technocrats themselves are seldom united.[4] Every class of federal experts has ambitious rivals. Every bureaucracy has counterbureaucracies staking disputed terrain.

Beneath that factionalism looms an angry critique of army construction sounding a familiar refrain. A denunciation of pork-barrel spending and environmental destruction, the critique puts the engineers between oil companies and Wall Street in our spook-house of traditional scapegoats. The Corps, say its defenders, suffers for sins of Congress, but engineers invite the abuse by overselling their science and lavishing public money on four-color books and pamphlets that downplay the long resistance to federal projects. There was seldom a time in American history, not even wartime, when the Corps worked in a vacuum without facing stiff opposition from rival organizations and bureaus.[5]

The Corps in the face of that protest seeks refuge in some ancient ideas: the belief that the truly scientific builder is the consummate action-oriented "all-around man," a generalist; also the notion that river building, vital to defense, keeps the agency poised for combat and ready for a domestic crisis, whether a chemical spill or a flood. "Our water resources mission and our work for other agencies help keep us trained and flexible, and directly support our mobilization mission," said Lt. Gen. Elvin R. Heiberg III, a Reagan-era chief engineer.[6] Here was the "army" in army construction, the Corps' combat mystique.

Reverence for that army scientific tradition may be a key to understanding the water agency that remains the nation's closet equivalent to a Corps des ponts et chaussées. In many ways the social-scientific ideals of the army-trained French engineers are a source of agency power. Faith in problem-solving through quantification, the view of rivers and harbors as technological systems, the commitment to social progress through strong government, the term "public works" itself—these are Enlightenment concepts, values that long predate the need to reclaim the desert or protect the Delta from floods. Thus the Corps can shift with the nation when its rivals sometimes cannot. While the Bureau of Reclamation is still pushing dams, while the New Deal's TVA, in mid-life crisis, loses its reformist appeal, the Corps, like the French engineers, moves from forts to rivers to most any colossal endeavor without straying far from its past.

In the end there is no simple way to characterize Corps engineering. Cautious, responsive, despotic, opportunistic—the engineers, after two hundred years, have learned to straddle the contradictions built into our nation-state. Corps planning promotes system and order. Corps field operations, locally implemented, serve a divided Congress rooted in community power. Can local self-government thrive under Corps supervision? Can environmentalists tolerate Corps expertise? These are historical quandaries no science will likely resolve.

Serious historians of federal water policy will search among the correspondence, diaries, journals, field surveys, and unpublished reports available in
the National Archives at the Old Army Branch, Record Group 77. These
records should be approached with the help of the Preliminary Inventory of
the Textual Records of the Office of the Chief of Engineers, part 1, compiled
by Elizabeth Bethel and revised by Maizie H. Johnson. A large body of correspondence relevant to the Corps' civil operations is found in Letters Sent
to Engineer Officers, 1812–1869, entry 6; and Letters Received, 1826–1866,
entry 18. For correspondence of the topographical bureau, see Letters Sent
to Others Than the Secretary of War, 1829–1867, entry 310; and Letters Received, 1832–1865, entry 315. The Bulky Package File, entry 292A, contains
unpublished reports and journals. Activities of the Board of Engineers for
Internal Improvements can be followed through a collection of General Macomb's correspondence in Letters Sent Relating to Internal Improvement,
1824–1830, entry 249. A General Information Index to Names and Subjects,
1789–1889, better known as the De Grange Index, entry 291, is the place to
begin tracking specific topics and events.

Important collections outside the Washington, D.C., area include the
Jonathan Williams Manuscripts, Lilly Library, Bloomington, Indiana; and
the Papers of Thomas Hutchins, Geographer General of the United States,
in the Historical Society of Pennsylvania, Philadelphia. The archives of the
American Philosophical Society, also in Philadelphia, house additional correspondence by Williams, Hutchins, and other people of science. Military science broadly defined is the subject of the Minutes and Records of the

United States Military Philosophical Society in the New York Historical Society in New York City. Perhaps the largest and best-indexed collection of manuscripts concerning the antebellum Corps is maintained in the library special collections of the United States Military Academy at West Point, New York.

Annual reports of the chief engineer (1823 to the present) and the chief topographical engineer (1831 to 1863) are a vast library on rivers and harbors. Nineteenth-century reports were published in the secretary of war's annual report to the president and bound with the papers of Congress in the congresssional serial set. For excerpts of selected reports before 1837, see *American State Papers, Class V, Military Affairs*, 7 vols. (Washington, D.C., 1832–1861). The legislative debates are found in *Annals of the Congress of the United States, 1789–1825* (Washington, D.C., 1834–1856) and *Congressional Globe, 1834–1873* (Washington, D.C., 1834–1873); see also James D. Richardson, comp., *A Compilation of the Messages and Papers of the Presidents, 1789–1902,* 10 vols. (Washington, D.C., 1896–1904), and U.S. Treasury, "Statement of the Appropriations and Expenditures for Public Buildings, Rivers and Harbors, Forts, Arsenals, Armories, and Other Public Works, from March 4, 1789, to June 30, 1882," S. Exec. Doc. 196, 47th Cong., 1st sess. (1882). This study also relied on the public speeches and petitions reprinted in *Niles Weekly Register* (Baltimore, 1811–1849) and *De Bow's Commercial Review of the South and the West* (New Orleans, 1846–1870).

Given the ocean of primary records, it is surprising that historians have written very little about the hot political issue of federal aid to rivers and harbors. During the 1950s, however, several good books were written about government and internal improvements. Forest G. Hill, in *Road, Rails, and Waterways: The Army Engineers and Early Transportation* (Norman, Okla. 1957), especially good on the U.S. Board of Engineers for Internal Improvement, says the federal government used engineering and science to open the West and stimulate private enterprise. This theme is also developed in W. Turrentine Jackson, *Wagon Roads West: A Study of Federal Road Surveys and Construction in the Trans-Mississippi West, 1846–1869* (Berkeley, 1952); William H. Goetzmann, *Army Exploration in the American West, 1803–1863* (New Haven, 1959); and Carter Goodrich, *Government Promotion of American Canals and Railroads, 1800–1890* (New York, 1960). George Rogers Taylor's seminal book, *The Transportation Revolution, 1815–1860* (New York, 1951), shows how publicly financed canals, roads, and railroads broke down the laissez-faire maritime orientation of "merchant capitalism," planting the seed of a vigorous, western-looking "national economy." Caroline E. MacGill et al. provide an encyclopedic review of antebellum projects in *History of*

Transportation in the United States before 1860 (Washington, D.C., 1917); see also Alvin F. Harlow, *Old Towpaths: The Story of the American Canal Era* (New York, 1926), and Ronald E. Shaw, *Canals for a Nation: The Canal Era in the United States, 1790–1860* (Lexington, Ky., 1990). On federal aid to the Ohio and Mississippi, Louis C. Hunter's *Steamboats on the Western Rivers: An Economic and Technological History* (Cambridge, Mass., 1949) remains in a class by itself.

Although no single volume tells the full political story of antebellum internal improvements, focused studies explain why the issue was so divisive. John L. Larson investigates the Jeffersonian-Madisonian position in "'Bind the Republic Together': The National Union and the Struggle for a System of Internal Improvements," *Journal of American History* 74 (1987): 363–387. See also Larson, "A Bridge, a Dam, a River: Liberty and Innovation in the Early Republic," *Journal of the Early Republic* 7 (Winter 1987): 351–375; Joseph H. Harrison, Jr., "*Sic et Non:* Thomas Jefferson and Internal Improvement," *Journal of the Early Republic* 7 (Winter 1987): 335–349; Harrison, "The Internal Improvement Issue in the Politics of the Union, 1783–1825" (Ph.D. diss., University of Virginia, 1954); Edmund C. Nelson, "Presidential Influence on the Policy of Internal Improvements," *Iowa Journal of History and Politics* 4 (January 1906): 3–69; R. B. Way, "The Mississippi Valley and Internal Improvements, 1825–1840," *Proceedings of the Mississippi Valley Historical Association* 4 (1910–1911): 153–180; Isaac Lippincott, "A History of River Improvement," *Journal of Political Economy* 22 (1914): 630–660; Harry N. Scheiber, "Federalism and the American Economic Order, 1789–1910," *Law and Society Review* 10 (Fall 1975): 58–118, and Donald J. Pisani, "Promotion and Regulation: Constitutionalism and the American Economy," *Journal of American History* 74 (December 1987): 740–768. For the fickle Jacksonians, see Raymond H. Pulley, "Andrew Jackson and Federal Support of Internal Improvements: A Reappraisal," *Essays in History* 9 (1963–1964): 48–59. Mentor L. Williams studies the Whig reaction to the Polk waterway vetoes in "The Chicago River and Harbor Convention, 1847," *Mississippi Valley Historical Review* 35 (March 1949): 607–626.

The era of Reconstruction revived the debate over government planning and waste. Edward Lawrence Pross's "History of Rivers and Harbors Appropriation Bills, 1866–1933" (Ph.D. diss., Ohio State University, 1938) shows how the Congress eagerly financed the postbellum construction boom. Albert Bushnell Hart dissects the appropriation process in a study of congressional logrolling; see "Biography of a River and Harbor Bill," *Papers of the American Historical Association* 3 (1888): 180–195. C. Vann Woodward explains why agrarians backed rivers and harbors appropriations in *Origins of the New South, 1877–1913* (Baton Rouge, 1951); see also Howard B. Schonberger, *Transportation to the Seaboard: The "Communication Revolution" and American Foreign Policy, 1860–1900* (Westport, Conn., 1971).

General histories of internal improvement say little about the ideology of science and engineering that shaped government from within. That understanding begins with William Goetzmann's *Exploration and Empire: The Explorer and the Scientist in the Winning of the American West* (New York, 1966). Stressing the subjective nature of scientific inquiry, Goetzmann shows how the works of government explorers and others were shaped by a number of factors—profession, politics, the idea of the West as a desert or as a garden, and the orthodoxy of science itself. No book has payed equal attention to the ideology of army construction, but historians have explored the values and methods of science that influenced the civilian profession, weaning it away from the crafts. See Edwin T. Layton, Jr., *The Revolt of the Engineers: Social Responsibility and the American Engineering Profession* (Baltimore, 1971); Layton, "Mirror Image Twins: The Communities of Science and Technology in Nineteenth-Century America," *Technology and Culture* 12 (October 1971): 562–580; Layton, "American Ideologies of Science and Engineering," *Technology and Culture* 17 (October 1976): 688–701; Max Lerner, "Big Technology and Neutral Technicians," *American Quarterly* 4 (Summer 1952): 99–109; Monte A. Calvert, *The Mechnical Engineer in America, 1830–1910: Professional Cultures in Conflict* (Baltimore, 1967); Bruce Sinclair, *Philadelphia's Philosopher Mechanics: A History of the Franklin Institute, 1824–1865* (Baltimore, 1974); and Robert V. Bruce, *The Launching of Modern American Science, 1846–1876* (New York, 1987).

The scientific ideology that shaped the Corps and West Point evolved from French engineering. Charles Coulston Gillispie's monumental work on *Science and Polity in France at the End of the Old Regime* (Princeton, 1980) discusses hydraulics, fort building, and the corps des ponts et chaussées. See also Frederick B. Artz, *The Development of Technical Education in France, 1500–1850* (Cambridge, Mass., 1966); Charles Stuart Gillmore, *Coulomb and the Evolution of Physics and Engineering in Eighteenth-Century France* (Princeton, 1971); Josef Konvitz, *Cartography in France, 1660–1848: Science, Engineering, and Statecraft* (Chicago, 1987); Turner Warton Allen, "The Highway and Canal System in Eighteenth Century France" (Ph.D. diss., University of Kentucky, 1953); André E. Guillerme, *The Age of Water: The Urban Environment in the North of France, A.D. 300–1800* (College Station, Tex., 1988); and Cecil O. Smith, Jr., "The Longest Run: Public Engineering and Planning in France," *American Historical Review* 95 (June 1990): 657–962. Good histories of the French influence on American technology include Roger G. Kennedy, *Orders from France: The Americans and the French in a Revolutionary World* (New York, 1989), and Peter Michael Molloy, "Technical Education and the Young Republic: West Point as America's *école Polytechnique, 1802–1833* (Ph.D. dissertation, Brown University, 1975).

The literature on eighteenth- and early nineteenth-century British engineering features gifted mechanics. Samuel Smiles, *Lives of the Engineers*, 3 vols. (London, 1862), covers Brindley, Smeaton, and others; see also E. C. R. Hadfield, *British Canals* (London, 1950), L. T. C. Rolt, *Navigable Waterways* (London, 1969), and Anthony Burton, *The Canal Builders* (Newton Abbot, 1981). W. G. H. Armytage's *Social History of Engineering* (Cambridge, Mass., 1961) compares British and French origins of the building professions. T. S. Willan's *River Navigation in England, 1600–1750* (London 1964) explains why the British builders came to prefer canals. For training and technical orienation, see A. E. Musson and Eric Robinson, *Science and Technology in the Industrial Revolution* (Toronto, 1969); see also Musson and Robinson, "The Origins of Engineering in Lancashire," *Journal of Economic History* 20 (1960): 209–230. Terry S. Reynolds explains how Smeaton used science; see "Scientific Influences on Technology: The Case of the Overshot Waterwheel, 1752–1754," *Technology and Culture* 20 (April 1979): 270–295; for national approaches to engineering science, see also Reynolds, *Stronger than a Hundred Men: A History of the Vertical Waterwheel* (Baltimore, 1983). David Landes says builders helped unleash forces of industrialization; see *The Unbound Prometheus: Technological Change and Industrial Development in Western Europe from 1750 to the Present Day* (Cambridge, Mass., 1969).

Waterways treatises suggest the importance of patriotism and national culture during a time of intense competition between builders in Britain and France. See, for example, Bernard Forest de Belidor, *Architecture Hydraulique; ou, L'art de conduire, d'élever, et de ménager les eaux pour les différens besoins de la vie . . .* , 2 vols. (Paris, 1737–1739); Pierre Louis Du Buat, *Principes d'Hydraulique; Ouvrage dans Lequel on Traite du Mouvement de l'Eau dans les rivières, canaux, etc.* (Paris, 1777); Charles Vallancey, *A Treatise on Inland Navigation* (Dublin, 1763); Joseph Banks, ed., *Reports of the Late John Smeaton*, 4 vols. (London, 1812); and John Phillips, *A General History of Inland Navigation* (London, 1805). For an American look at European canals, see Robert Fulton, *A Treatise on the Improvement of Canal Navigation* (London, 1796). and Robert Mills, *A Treatise on Inland Navigation* (Baltimore, 1829).

Studies of early American engineering emphasize the word-of-mouth, trial-and-error tradition of the craftsman-mechanic. The standard reference remains Daniel Hovey Calhoun's work, *The American Civil Engineer: Origins and Conflict* (Cambridge, Mass., 1960), a good study of professionalization that examines the shortage of competent builders before the Erie Canal. Darwin H. Stapleton explores the contribution of William Weston and the largely British orgins of American civil engineering in *The Transfer of Early Industrial Technologies to America* (Philadelphia, 1987). See also Stapleton, ed., *The Engineering Drawings of Benjamin Henry Latrobe* (New Haven, Conn., 1980); Melvin Kranzberg, "Transferring a Dynamic Technology during Socio-Economic Change: The American Experience, 1607–1850," in Colloques Internationaux de Centre National de la Recherche Scientifique, *L'Acquisition des Techniques par les Pays Non-Initiateurs* (Paris, 1973); and Elting E. Morison, *From Know-How to Nowhere: The Development of American Technology* (New York, 1974). Two articles by Eugene S. Ferguson show how mechanics and inventors mixed British and American traits; see "On the Origin and Development of American Mechanical 'Know-How,'" *Midcontinent American Studies Journal* 3 (1962): 3–5, and "The American-ness of American Technology," *Technology and Culture* 20 (January 1979): 3–24. Silvio A. Bedini looks at surveyors and scientific instruments in *Tinkers and Thinkers: Early American Men of Science* (New York, 1975). Colorful European accounts of American public works include Michael Chevalier's *Society, Manners, and Politics in the United States* (Boston, 1839) and David Stevenson's *Sketch of the Civil Engineering of North America* (London, 1838).

Histories of the Corps and West Point show the endurance of European ideas. See Stephen E. Ambrose, *Duty, Honor, Country: A History of West Point* (Baltimore, 1966); Edward Burr, *Historical Sketch of the Corps of Engineers, U.S. Army* (Washington, D.C., 1939); Albert E. Cowdrey, *A City for the Nation: The Army Engineers and the Building of Washington, D.C., 1798–1967* (Washington, D.C., 1979); George W. Cullum, *Campaigns of the War of 1812–15 against Great Britain* (New York, 1879); and Paul K. Walker, *Engineers of Independence: A Documentary History of the Army Engineers in the American Revolution, 1775–1783* (Washington, D.C., 1981). Martin Reuss probes the military-civilian conflict and the Corps' organizational culture in "Andrew A. Humphreys and the Development of Hydraulic Engineering: Politics and Technology in the Army Corps of Engineers, 1850–1950," *Technology and Culture* 26 (January 1985): 1–33. Garry D. Ryan, "War Department Topographical Bureau, 1831–1863" (Ph.D. diss., American University, 1968), is essential reading for students of army bureaucracy and administration.

Corps district and division histories are an excellent source for details about early projects. Good histories of antebellum construction include

Leland R. Johnson, *The Falls City Engineers: A History of the Louisville District, Corps of Engineers, United States Army* (Louisville, Ky., 1974); Johnson, *The Davis Island Lock and Dam, 1870–1922* (Pittsburgh, 1985); Frank E. Snyder and Brian H. Guss, *The District: A History of the Philadelphia District, U.S. Army Corps of Engineers, 1866–1971* (Philadelphia, 1974); Aubrey Parkman, *Army Engineers in New England: The Military and Civil Works of the Corps of Engineers in New England, 1775–1975* (Waltham, Mass., 1978); and Gary B. Mills, *Of Men and Rivers: The Story of the Vicksburg District* (Vicksburg, Miss., 1978). Walter M. Lowrey includes Corps projects in a superb dissertation, "Navigational Problems at the Mouth of the Mississippi River, 1698–1880" (Vanderbilt University, 1956).

The history of the Corps is also military history, although few books on the Old Army touch on engineering ideology and politics. Two rewarding exceptions are Theodore J. Crackel, *Mr. Jefferson's Army: Political and Social Reform of the Military Establishment, 1801–1809* (New York, 1987), and Philip L. Shiman, "Engineering Sherman's March: Army Engineers and the Management of Modern War" (Ph.D. diss., Duke University, 1991). Because so many engineers went on to fame as generals, explorers, and politicians, there are many excellent biographies, such as Stephen W. Sears, *George B. McClellan: The Young Napoleon* (New York, 1988); Joseph G. Tregle, introduction to Thomas Hutchins, *An Historical Narrative and Topographical Description of Louisiana and West-Florida* (Gainesville, Fla., 1968), pp. v–xlviii; Russell F. Weigley, *Quartermaster General of the Union Army: A Biography of M. C. Meigs* (New York, 1959); and Richard G. Wood, *Stephen Harriman Long, 1784–1865: Army Engineer, Explorer, Inventor* (Glendale, Calif., 1966).

Finally there are those articles and books that say nothing about antebellum construction but a great deal about how technologists and planners influence the public debate. Six studies deserve special mention: Rue Bucher and Anselm Strauss, "Professions in Process," *American Journal of Sociology* 66 (January 1961): 325–344; John A. Ferejohn, *Pork Barrel Politics: River and Harbors Legislation, 1947–1968* (Stanford, Calif., 1974); Otis L. Graham, Jr., *Toward a Planned Society: From Roosevelt to Nixon* (New York, 1976); Samuel P. Hayes, *Conservation and the Gospel of Efficiency: The Progressive Conservation Movement, 1890–1920* (Cambridge, Mass., 1959); Arthur Maass, *Muddy Waters: The Army Engineers and the Nation's Rivers* (Cambridge, Mass., 1951); and Leonard A. Shabman, *Decision Making in Water Resources Investment and Potential of Multi-Obligation Planning: The Case of the Army Corps of Engineers* (Ithaca, N.Y., 1972).

Notes

Prologue: A Nation Builder

1. Henry G. Wheeler, *History of Congress, Biographical and Political* (New York: Harper & Brothers, 1848), pp. 305, 339–340.

2. For the Corps as a "bureaucratic superstar," see Jeanne Nienaber Clarke and Daniel McCool, *Staking Out the Terrain: Power Differentials among Natural Resource Management Agencies* (Albany: State University of New York Press, 1985), pp. 13–33; for the "preeminent" Corps, see Stephen Klaidman, "Engineer Corps: Benefits Contested," *Washington Post,* July 6, 1975; see also George Laycock, *The Diligent Destroyers* (New York: Doubleday & Co., 1970); Robert de Roos and Arthur A. Maass, "The Lobby That Can't Be Licked: Congress and the Army Engineers," *Harper's* 199 (August 1974): 21–23; and William O. Douglas, "The Public Be Damned," *Playboy* 16 (July 1969): 143, 182, 184–188.

3. Williams to Decius Wadsworth, August 13, 1802, Letterbooks, January 15, 1802–December 28, 1806, Jonathan William Manuscripts, Lilly Library, Indiana University, Bloomington.

4. Melvin Kranzberg, "The Disunity of Science-Technology," *American Scientists* 56 (1968): 21. Otto Mayr believed that modern academic distinctions between "science" and "technology" were confusing and perhaps less important than case studies that seek to explain how people in the past were actually using those slippery terms; see "The Science-Technology Relationship as a Historiographical Problem," *Technology and Culture* 17 (January 1976): 663–671.

5. Albert Bushnell Hart, "Biography of a River and Harbor Bill," *Papers of the American Historical Association* 3 (1888): 183.

6. *The Nation* 69 (February 9, 1899): 100; for Congressman Burton's reforms and the river act of 1899, see Edward Lawrence Pross, "A History of Rivers and Harbors Appropriation Bills" (Ph.D. diss., Ohio State University, 1938), pp. 104–105.

7. George Y. Wisner, "Worthless Government Engineering," *Engineering Magazine* 2 (January 1892): 433–434.

8. John L. Mathews, "The Future of Our Navigable Rivers," *Atlantic Monthly* 100 (December 1907): 724.

9. David E. Wilson, *The National Planning Idea in U.S. Policy: Five Alternative Approaches* (Boulder, Colo.: Westview Press, 1980), pp. 1–2; for the Corps' resistance to planning, see M. O. Leighton, "The Relation of Water Conservation to Flood Prevention and Navigation in the Ohio River," *Engineering News* 59 (May 7, 1908): 498–504.

10. Otis L. Graham, Jr., *Toward a Planned Society: From Roosevelt to Nixon* (New York: Oxford University Press, 1976), p. 56.

11. Elmo Richardson, *Dams, Parks and Politics: Resource Development and Preservation in the Truman-Eisenhower Era* (Lexington, Ky.: University of Kentucky Press, 1973), pp. 25, 47, passim; see also Samuel P. Hays, *Conservation and the Gospel of Efficiency: The Progressive Conservation Movement, 1890–1920* (Cambridge: Harvard University Press, 1959).

12. Harold L. Ickes's foreword to Arthur Maass, *Muddy Waters: The Army Engineers and the Nation's Rivers* (Cambridge: Harvard University Press, 1951), p. ix.

13. Arthur E. Morgan, *Dams and Other Disasters: A Century of Army Engineering in Civil Works* (Boston: Porter Sargent, 1971), pp. 2, 6.

14. Fred Powledge, *Water: The Nature, Uses, and Future of Our Most Precious and Abused Resource* (New York: Farrar Straus Giroux, 1982), p. 298.

15. Marc Reisner, *Cadillac Desert: The American West and Its Disappearing Water* (New York: Viking, 1986), p. 178; for the Corps as a technocratic elite, see also K. Bruce Galloway and Robert Bowie Johnson, Jr., *West Point: America's Power Fraternity* (New York: Simon and Schuster, 1973), pp. 221–224.

16. Everett Mendelsohn, Merritt Roe Smith, and Peter Weingart, eds., *Science, Technology, and the Military*, 2 vols. (Dordrecht, Netherlands: Kluwer Academic Publishers, 1988), 1:xv.

17. *Engineer Memoirs: Interviews with Lieutenant General Frederick J. Clarke*, Engineer Pamphlet 870-1-5 (Washington, D.C.: Historical Division, Office of the Chief Engineer, 1979), p. 207.

18. Karl W. Deutsch, *Politics and Government: How People Decide Their Fate* (Boston: Houghton Mifflin Co., 1970), p. 192; see also Francis E. Rourke, *Bureaucracy, Politics, and Public Policy* (Boston: Little, Brown & Co., 1976), p. 18.

19. Captain William H. Bixby, "Discussion on the South Pass Jetties—Ten Years of Practical Teachings in River and Harbor Hydraulics," *Transactions*, American Society of Civil Engineers, 15 (April 1886): 256.

20. Jacob M. Dickinson, "The Army Engineer and River Improvement," *Professional Memoirs* 2 (January–March 1910): 73.

21. Samuel C. Florman, "Hired Scapegoats," *Harper's* 254 (May 1977): 28.

22. Leonard A. Shabman, *Decision Making in Water Resource Investment and Potential of Multi-Obligation Planning: The Case of the Army Corps of Engineers*, Technical Report 42 (Ithaca, N.Y.: Cornell University Water Resource and Marine Science Center, 1972), p. 101; for the argument that "the Corps of Engineers does merely what is assigned to it," see W. Stull Holt, *The Office of the Chief Engineers of the Army: Its Non-Military History, Activities, and Organization* (Baltimore: Johns Hopkins University Press, 1923), p. 32.

23. Alex Roland, "Technology and War: A Bibliographical Essay," in Merritt Roe Smith, ed., *Military Enterprise and Technological Change: Perspectives on the American Experience* (Cambridge: MIT Press, 1985), p. 372.

24. Raymond H. Merritt, *Creativity, Conflict and Controversy: A History of the St. Paul District, U.S. Army Corps of Engineers* (Washington, D.C.: U.S. Government Printing Office, 1979), p. 40; see also Forest Hill, *Roads, Rails, and Waterways: The Army Engineers and Early Transportation* (Norman: University of Oklahoma Press, 1957), Ronald B. Hartzer, *To Great and Useful Purpose: A History of the Wilmington District, U.S. Army Corps of Engineers* (Wilmington, N.C.: U.S. Army Corps of Engineers, 1984), and Nuala Drescher, *Engineers for the Public Good: A History of the Buffalo District, U.S. Army Corps of Engineers* (Buffalo, N.Y.: U.S. Army Corps of Engineers, 1982); for the argument that "the science-military connection desperately needs case studies," see Mendelsohn, Smith, and Weingart, *Science, Technology, and the Military*, p. xx.

25. Charles F. O'Connell, Jr., "The Corps of Engineers and the Rise of Modern Management, 1827–1856," in Smith, *Military Enterprise and Technological Change*, pp. 87–116.

26. Critics of industrial capitalism often cite the military as a source of authoritarian values; see, for example, Lewis Mumford, *Technics and Civilization* (New York: Harcourt Brace and Company, 1934), pp. 89–96; for the army as a seed of civil engineering, see Sidney Forman, "The First School of Engineering," *Military Engineer* 44 (March April 1952): 109 112.

27. William F. Ogburn, *Living with Machines* (Chicago: American Library Association, 1933), p. 15.

1. European Antecedents

1. Latrobe to Robert Mills, July 12, 1806; reprinted in Talbot Hamlin, *Benjamin Henry Latrobe* (New York: Oxford University Press, 1955), pp. 585–591.

2. Latrobe to James Madison, April 8, 1816; reprinted with notes in Edward C. Carter II, *Benjamin Henry Latrobe and Public Works: Professionalism, Private Interest, and Public Policy in the Age of Jefferson*, Essays in Public Works History, no. 3 (Chicago: Public Works Historical Society, 1976), pp. 22–29.

3. Thomas Moore, *Ship Navigation to Georgetown* (n.p., 1811), p. 4.

4. Benjamin Henry Latrobe, *The Virginia Journals of Benjamin Henry Latrobe, 1795–1798*, ed. Edward C. Carter II, 2 vols. (New Haven: Yale University Press, for The Maryland Historical Society, 1977), 1:79.

5. Latrobe to Thomas More [Moore?], January 20, 1811; in Hamlin, *Benjamin Henry Latrobe*, p. 564.

6. Latrobe to James Madison, April 8, 1816; in Carter, *Latrobe and Public Works*, p. 19.

7. "On Inventions," *Useful Cabinet* 3 (March 1808): p. 58; for the British influence, see Darwin H. Stapleton, *The Transfer of Early Industrial Technologies to America* (Philadelphia: American Philosophical Society, 1987), pp. 1, 21–31.

8. Melvin Kranzberg, "Transferring a Dynamic Technology during Socio-Economic Change: The American Experience, 1607–1850," *L'Acquisition des Techniques par les Pays Non-Initiateurs*, Colloques Internationaux de Centre National de la Recherche Scientifique (Paris: Centre National de la Recherche Scientifique, 1973), p. 265.

9. Elting E. Morison, *From Know-How to Nowhere: The Development of American Technology* (New York: Basic Books, 1974), p. 90.

10. Eugene S. Ferguson, "On the Origin and Development of American Mechanical 'Know-How,'" *Midcontinent American Studies Journal* 3 (Fall 1962): 405–428.

11. The marquis de Condorcet's reference to the scientific philosophy of minister Anne-Robert-Jacques Turgot, quoted in Charles Coulston Gillispie, *Science and Polity in France at the End of the Old Regime* (Princeton: Princeton University Press, 1980), p. 3.

12. Henry Guerlac, "Vauban: The Impact of Science on War," in *Makers of Modern Strategy: Military Thought from Machiavelli to Hitler*, ed. Edward Mead Earle (Princeton: Princeton University Press, 1943), pp. 26–43; see also Josef Konvitz, *Cartography in France, 1660–1848* (Chicago: University of Chicago Press, 1987), p. 93.

13. Paul Gille, "Sea and River Transportation," in *A History of Technology and Invention: Progress through the Ages*, vol. 2, *The First Stages of Mechanization* (New York: Crown Publishing Group, 1969), pp. 402–404.

14. Turner Warton Allen, "The Highway and Canal System in Eighteenth Century France" (Ph.D. diss., University of Kentucky, 1953), pp. 678–686; see also Roger Price, *An Economic History of Modern France, 1730–1914* (New York: St. Martin's Press, 1975), pp. 12–20.

15. Charles Hadfield, *The Canal Age* (Newton Abbot, U.K.: David & Charles, 1968), pp. 176–177; for a full history, see T. C. Rolt, *From Sea to Sea: The Canal du Midi* (Athens: Ohio State University Press, 1974).

16. Franklin to S. Rhoads, mayor of Philadelphia, from London, August 22, 1772; quoted in Archer Butler Hulbert, *Historic Highways of America*, vol. 13, *The Great American Canals* (1904; reprint, New York: AMS, 1971), pp. 25–29.

17. Robert F. Hunter and Edwin L. Dooley, Jr., *Claudius Crozet: French Engineer in America, 1790–1864* (Charlottesville: University of Virginia Press, 1989), p. 39; see also Julian P. Boyd, ed., *The Papers of Thomas Jefferson*, 19 vols. (Princeton: Princeton University Press), 11:371–372.

18. Charles Woolsey Cole, *Colbert and a Century of French Mercantilism*, 2 vols. (Hamden, Conn.: Archon Books, 1964), 1:381.

19. Rolt, *From Sea to Sea*, p. 97; see also Allen, "Highway and Canal System in Eighteenth Century France," pp. 329–375, 510.

20. A. Blanchard, "Les Ci-Devant Ingénieurs du Roi," *Revue Internationale d'Histoire Militaire* 30 (1970): 97–108; see also Charles Stuart Gillmore, *Coulomb and the Evolution of Physics and Engineering in Eighteenth-Century France* (Princeton: Princeton University Press, 1971), pp. 9–18, and Carl B. Boyer, *A History of Mathematics* (Princeton: Princeton University Press, 1968), pp. 511–512, 573.

21. Gillispie, *Science and Polity*, pp. 479–498; for the importance of mathematics, see Frederick B. Artz, *The Development of Technical Education in France* (Cambridge: MIT Press, 1966), p. 68, passim.

22. The Library of Congress's four-book set of Belidor's masterpiece has a note on the second volume: "Paris 1737: One of the works purchased by Monroe when [visiting] France, 1794–1796, and included in his private collection"; see Bernard Forest de Belidor, *Architecture Hydraulique; ou, L'art de conduire, d'élever, et de ménager les eaux pour les différens besoins de la vie . . .* , 2 vols. (Paris: C. Jombert, 1737–1739).

23. Andrée Guillerme, *Age of Water: The Urban Environment in the North of France, A.D. 300–1800* (College Station: Texas A&M University Press, 1988), p. 195.

24. Denis Diderot, comp., *Encyclopaedia,* 1751–1765 ed., s.v. "Hydraulique" and "Hydrodynamique"; see also Hunter Rouse and Simon Ince, *History of Hydraulics* (New York: Dover Publishing, 1957), pp. 113–138.

25. Guglielmini's *Opera Omnia Mathematica, hydraulica, medica, et physica* appeared posthumously in 1719.

26. Guillerme, *Age of Water,* p. 199; see also Charles Vallancey, *A Treatise on Inland Navigation* (Dublin: George and Alexander Ewins, 1763), pp. 10–12, and Richard S. Westfall, "Floods along the Bisenzio: Science and Technology in the Age of Galileo," *Technology and Culture* 30 (October 1989): 905–906.

27. Robert Mills, *A Treatise on Inland Navigation* (Baltimore: Lucas, 1829), p. 19; for problems with French river and canal engineering, see Rick Szostak, *The Role of Transportation in the Industrial Revolution: A Comparison of England and France* (Montreal and Kingston: McGill-Queen's University Press, 1991), pp. 54–60.

28. Charles Coulston Gillispie, *Lazare Carnot Savant* (Princeton: Princeton University Press, 1971), pp. 20–21, 24–25.

29. Terry Shinn, *Savoir Scientifique et Pouvoir Sociale: L'Ecole Polytechnique, 1794–1914* (Paris: Presses de la Fondation Nationale des Sciences Politiques, 1980), esp. ch. 1; see also Janis Langins, "The Ecole Polytechnique and the French Revolution: Merit, Militarization, and Mathematics" (Paper delivered at the Eighteenth International Congress of the History of Science, August 1–9, 1989, Hamburg and Munich, Germany), and Margaret Bradley, "Scientific Education versus Military Training: The Influence of Napoleon Bonaparte on the *Ecole Polytechnique,*" *Annals of Science* 32 (1975): 415–449. Presumably Professor Joseph-Marie Sganzin was the same engineer as the French inspector general M. I. Sganzin whose notes were translated for West Point cadets in 1823; see M. I. Sganzin, *An Elementary Course of Civil Engineering,* translated from the 3rd French ed. (Boston: Hilliard, Gray, Little, and Wilkins, 1827).

30. Cecil O. Smith, "The Longest Run: Public Engineers and Planning in France, *American Historical Review* 95 (June 1990): 660.

31. Guillerme, *Age of Water,* p. 206.

32. L. T. C. Rolt, *Navigable Waterways* (London: Longmans, 1969), pp. 3–10; for British improvements before the Bridgewater Canal see Thomas S. Willan, *River Navigation in England, 1600–1750* (New York: August M. Kelley, 1936), and A. W. Skempton, "The Engineers of the English River Navigations, 1620–1760," *Transactions,* Newcomen Society, 29 (1953): 25; for bear-trap dams, see Howard Stansbury, "Survey of the Cumberland River," H. Doc. 171, 23d Cong., 1 sess. (1835), pp. 8–11.

33. Vallancey, *Treatise on Inland Navigation,* pp. iv–v; see also Douglas William Marshall, "The British Military Engineer, 1741–1783: A Study of Organization, Social Origin, and Cartography" (Ph.D. diss., University of Michigan, 1976), pp. 49, 85–86.

34. Sir Bosdin Leech, *History of the Manchester Ship Canal,* 2 vols. (Manchester: Sherrat & Hughes, 1907), 1:17.

35. John Phillips, *A General History of Inland Navigation* (London: B. Crosby & Co., 1805), p. viii.

36. Anthony Burton, *The Canal Builders* (London: David & Charles, 1981), pp. 12–18; see also Phyllis Deane, *The First Industrial Revolution* (Cambridge: Oxford University Press, 1965), p. 72–79; and Joseph Priestley, *Historical Account of the Navigable Rivers, Canals, and Railways of Great Britain* (London: Longmans, 1831), pp. vii-xiv.

37. Robert Fulton, *A Treatise on the Improvement of Canal Navigation* (London: I & J Tayler, 1796), p. 9.

38. J. Walker, *British Economic and Social History, 1700–1967* (London: MacDonald & Evans, 1968), p. 78.

39. Hulbert, *Historic Highways of America* 13:26–27.

40. John Phillips, quoted in Burton, *Canal Builders*, p. 82.

41. Samuel Smiles, *Lives of the Engineers*, 3 vols. (New York: Augustus M. Kelley, 1968), 1:471.

42. Cyril T. G. Boucher, *James Brindley: Engineer* (Norwich: Goose and Son, 1968), pp. 39–45; see also Lawrence Meynell, *James Brindley: The Pioneer of Canals* (London: Werner Laurie, 1956), pp. 38–46.

43. Smiles, *Lives of the Engineers* 1:473–474.

44. *Encyclopaedia Britannica*, 1771 ed., s.v. "Engineer."

45. Joseph Banks, ed., *Reports of the Late John Smeaton*, 4 vols. (London: Longmans, 1812), 1:v.

46. S. B. Donkin, "The Society of Civil Engineers," *Transactions*, Newcomen Society 17 (1936–1937): 57–71; for the influence of European hydraulics, see *John Smeaton's Diary of His Journey to the Low Countries, 1755* (Leamington: Courier Press, 1938), p. 155.

47. William Fairbairn, *A Treatise on Mills and Millwork*, 2 vols. (London, 1861–1862), 1:v.

48. John Smeaton, "Experimental Enquiry," *Philosophical Transactions of the Royal Society of London* 59 (1759), reprinted as a pamphlet the following year; see also Terry S. Reynolds, "Scientific Influences on Technology: The Case of the Overshot Waterwheel, 1752–1754," *Technology and Culture* 20 (April 1979): 270–295, and A. E. Musson and Eric Robinson, "The Origins of Engineering in Lancashire," *Journal of Economic History* 20 (1960): 209–230.

49. Stephen P. Timoshenko, *History of Strength of Materials* (New York: McGraw-Hill, 1953), p. 100.

50. Edward Deering Mansfield, "United States Military Academy at West Point," *American Journal of Education* 30 (March 1863): 17, 29, 42–43.

51. L. T. C. Rolt, *Thomas Telford* (London: Longmans, Green, & Co., 1958), p. 191.

52. W. H. G. Armytage, *The Rise of the Technocrats: A Social History* (London: Routledge and Kegan Paul, 1965), pp. 134–137.

53. Washington to the president of Congress, July 10, 1775, in John C. Fitzpatrick, ed., *The Writings of George Washington*, 30 vols. (Washington, D.C.: U.S. Government Printing Office, 1931–1944) 3:322, 4:196; see also Raleigh B. Buzzaird, "America's First Chief Engineer," *Military Engineer* 39 (1947): 404–510.

54. Rowena Buell, ed., *The Memoirs of Rufus Putnam* (Boston: Houghton, Mifflin, 1903), p. 55.

55. Darwin H. Stapleton, "Moncure Robinson: Railroad Engineer, 1828–1840," in *Benjamin Henry Latrobe and Moncure Robinson: The Engineer as Agent of Technological Transfer*, ed. Barbara E. Benson, Regional Conference in Economic History, May 17, 1974 (Wilmington, Del.: Eleutherian Mills-Hasley Foundation, 1975), p. 33.

56. Paul K. Walker, *Engineers of Independence: A Documentary History of the Army Engineers in the American Revolution, 1775–1783* (Washington, D.C.: Historical Division, Office of the Chief Engineer, 1981), p. 150.

57. Philippe Tronson du Coudray, "Coudray's Observations on Forts for Defense of the Delaware, July 1977, *Pennsylvania Magazine of History and Biography* 24 (1900): 343–347; see also Raleigh B. Buzzaird, "Washington's Favorite Engineer," *Military Engineer* 40 (1948): 115–118; and, William F. Heavay, "The Corps in the Days of the Revolution," *Military Engineer* 31 (November–December 1939): 411.

58. Howard Mumford Jones, *America and French Culture, 1750–1848* (Chapel Hill: University of North Carolina, 1927), pp. 124, 126.

59. Some officers were French Canadians; see Francis B. Heitman, *Historical Register of the Officers of the Continental Army during the War of the Revolution* (Baltimore: Genealogical Publishing Co., 1973), p. 642; for numbers of French soldiers and sailors, see Jones, *America and French Culture*, p. 524; see also Walker, *Engineers of Independence*, p. 17; for examples of French cartography during the War of Independence, see Howard C. Rice and Anne S. Brown, eds. and trans., *The American Campaigns of Rochambeau's Army, 1780, 1781, 1782, 1783*, 2 vols. (Princeton: Princeton University Press, 1972).

60. Duportail vilified America in a letter to Saint Germain, France's minister of war: "There is a hundred times more enthusiasm for this revolution in any café in Paris than in all the United States together"; see Durand Echeverria, *Mirage in the West: A History of the French Image of American Society to 1815* (New York: Octagon Books, 1966), p. 84.

61. Walker, *Engineers of Independence*, pp. 21, 349, 352.

62. Washington to president of Congress, October 31, 1781, in Elizabeth S. Kite, *Brigadier-General Louis Lebeque Duportail* (Baltimore: Johns Hopkins University Press, 1933), p. 220.

63. Jean-Baptiste Duroselle, *France and the United States: From the Beginnings to the Present Day*. Trans. Derek Coltman. (Chicago: University of Chicago, 1978), p. 23.

64. Walker, *Engineers of Independence*, pp. 330–331.

65. Ibid., pp. 356–358.

66. Leland Johnson, "The Fourth Pillar of Defense: Waterways," *National Waterways Roundtable Papers*, National Waterways Study, U.S. Army Engineer Water Resources Support Center, Institute for Water Resources (Washington, D.C.: U.S. Government Printing Office, 1980), pp. 23–35; see also W. Wright, "Notes of the Siege of Yorktown in 1781 with Special Reference to the Conduct of a Siege in the Eighteenth Century," *William and Mary College Quarterly*, 2d series, vol. 12 (1939), pp. 229–249; and H. Paul Caemmerer, *The Life of Pierre Charles L'Enfant* (Washington: National Republic Publishing Co., 1950), pp. 1–24.

67. Michael Chevalier, *Society, Manners, and Politics in the United States*, (1839, reprint, New York: Burt Franklin, 1969), pp. 245–246.

68. David Stevenson, *Sketch of the Civil Engineering of North America* (London: John Weale, 1838), pp. 20, 27.

69. Armytage, *Rise of the Technocrats*, p. 136.

70. New York Canal Commissioners, *Report*, February 15, 1817, in *Laws of the State of New York, in Relation to the Erie and Champlain Canals, Together with Annual Reports of Canal Commissioners . . .*, 2 vols. (Albany: E. and E. Horsford, 1825), 1:117, 372, 380; see also Daniel Hovey Calhoun, *The American Civil Engineer: Origins and Conflict* (Cambridge: MIT Press, 1960), p. 27; for American building traits, see Eugene Ferguson, "The American-ness of American Technology," *Technology*

and Culture 20 (January 1979): 3–24; Darwin Stapleton, "The American Model for the Genesis of European railroads: The Role of Values Transfers of Technology" (Paper read to the Society for the History of Technology, December 1979 [typewritten]), pp. 3–5, 19, and Brooke Hindle, ed., *America's Wooden Age: Aspects of Its Early Technology* (Tarrytown, N.Y.: Sleepy Hollow Restorations, 1975), p. 9.

71. Washington to Edmund Randolph, September 16, 1785, quoted in Desmond FitzGerald, "Address [on early engineers] at the Annual Convention at Cape May," *Transactions,* American Society of Civil Engineers, 41 (June 1899): 598.

72. B. H. Latrobe to Christian Ignatius Latrobe, November 4, 1804, quoted in Carter, *Latrobe and Public Works,* p. 7.

73. For Weston's role as an agent of British construction, see Stapleton, *Transfer of Early Industrial Technologies,* pp. 50–59, 126; see also Richard Shelton Kirby, "William Weston and His Contribution to Early American Engineering," *Transactions,* Newcomen Society, 16 (1935–1936): 111–127; and Morison, *From Know-How to Nowhere,* pp. 23–37.

74. FitzGerald, "Address at Cape May," pp. 601–603; see also *Dictionary of American Biography,* 1936 ed. (hereafter cited as *DAB*), s.v. "Weston, William."

75. "Report of William Weston, Esq.," in U.S. Senate, *Reports and Draughts of Surveys for the Improvement of Harbors and Rivers and the Construction of Roads and Canals,* vol. 1, 1789–1823 (Washington, D.C.: Gales and Seaton, 1839), p. 411.

76. For the Anglo-American preference for canals over river locks, see Tobias Lear, *Observations of the River Potomack, Etc.* (Baltimore: Samuel T. Chambers, 1793), p. 20; see also Richard Wentworth's comments on Brindley's philosophy in Mills, *Treatise on Inland Navigation,* pp. 21–22, and Phillips, *History of Inland Navigation,* p. 229.

77. "Report of William Weston, Esq.," p. 411.

78. *North American Review* 8 (1819): 15.

79. Eugene S. Ferguson, "On the Origin and Development of American Mechanical 'Know-How,'" *Midcontinent American Studies Journal* 3 (Fall 1962): 405–428; see also Neal Fitzsimons, *The Reminiscences of John B. Jarvis, Engineer of the Old Croton* (Syracuse, N.Y.: Syracuse University, 1971), p. 40, and Evald Rink, *Technical Americana: A Checklist of Technical Publications Printed before 1831* (Millwood, N.Y.: Kraus International Publications, 1981), p. 283.

80. W. H. G. Armytage, *A Social History of Engineering* (London: MIT Press, 1961), pp. 122–123.

81. Anthony F. C. Wallace, *The Social Context of Innovation: Bureaucrats, Families, and Heroes in the Early Industrial Revolution, as Foreseen in Bacon's* New Atlantis (Princeton: Princeton University Press, 1982), p. 153.

82. For the work ethic, legal institutions, and other "non-economic assets" that can facilitate technology transfer, see David J. Jeremy, *Transatlantic Industrial Revolution: The Diffusion of Textile Technologies between Britain and America, 1790–1830s* (Cambridge: MIT Press, 1981), p. 257; see also Oscar Handlin, "International Migration and the Acquisition of New Skills," in Bert F. Hoselitz, ed., *The Progress of Underdeveloped Areas* (Chicago: University of Chicago Press, 1952), pp. 54–59.

83. Martin S. Pernick, "Politics, Parties, and Pestilence: Epidemic Yellow Fever in Philadelphia and the Rise of the First Party System," *William and Mary Quarterly,* 3d series, 29 (October 1972): 559–580; for the isolation of the French intellectual, see

Francis Sargeant Childs, *French Refugee Life in the United States, 1790–1800: An American Chapter of the French Revolution* (Philadelphia: Porcupine Press, 1978); see also Roger G. Kennedy, *Orders from France: The Americans and the French in a Revolutionary World, 1780–1820* (New York: Alfred A. Knopf, 1989), p. 185.

84. Hill, *Roads, Rails, and Waterways,* p. 210.

85. Calhoun, *American Civil Engineer,* p. 43.

86. O'Connell, "The Corps and Modern Management," pp. 87–116; for West Point as the parent of civil engineering, see Ellis L. Armstrong, ed., *History of Public Works in the United States* (Chicago: American Public Works Association, 1976), pp. 667–668.

87. Kennedy, *Orders from France,* pp. 53, 55.

88. Latrobe to S. Gordon, January 14, 1813, in Hamlin, *Benjamin Henry Latrobe,* p. 386.

89. Morgan, *Dams and Other Disasters,* p. 38.

90. Calhoun, *American Civil Engineer,* p. 20; for the artisan resistance to military innovation, see Merritt Roe Smith, *Harpers Ferry Armory and the New Technology: The Challenge of Change* (Ithaca, N.Y.: Cornell University Press, 1977), pp. 20–21.

2. Mapping Water, Marking Land

1. Henry A. S. Dearborn, *Letters on the Internal Improvements and Commerce of the West* (Boston: Dutton and Wentworth, 1839), pp. 51, 59–60.

2. U.S. Congress, *American State Papers, Military Affairs* (hereafter cited as *ASP, MA*), 7 vols. (Washington, D.C.: Gales and Seaton, 1832–1861), 1:431.

3. Garry David Ryan, "War Department Topographical Bureau, 1831–1863: An Administrative History" (Ph.D. diss., The American University, 1968), pp. 1–16; see also Edward Burr, "Historical Sketch of the Corps of Engineers, United States Army," *Occasional Papers, The Engineers School, United States Army* 71 (Washington, D.C., 1939): 499, and Henry P. Beers, "A History of the U.S. Topographical Engineers, 1813–1863," *Military Engineer* 34 (June 1942): 287–291. William H. Goetzmann's research on the topographical corps and its scientific mission is still unsurpassed; see his *Army Exploration in the American West, 1808–1863* (New Haven: Yale University Press, 1959) and *Exploration and Empire: The Explorer and the Scientist in the Winning of the American West* (New York: W. W. Norton, 1966), esp. pp. 58–64, 231–264.

4. W. Turrentine Jackson, *Wagon Roads West: A Study of Federal Road Surveys and Construction in the Trans-Mississippi West, 1846–1889* (Berkeley and Los Angeles: University of California Press, 1952), p. 2.

5. Henry C. Boger, Jr., "The Role of the Army Engineers in the Westward Movement in the Lake Huron Michigan Basin before the Civil War" (Ph.D. diss., Columbia University, 1954), p. 24.

6. Totten to Secretary of War Calhoun, February 8, 1824, Letters and Reports of Col. Joseph G. Totten, Chief of Engineers, 1803–1864, 10 vols., 1:17–23, entry 146, Textual Records of the Office of the Chief of Engineers (hereafter cited as Totten Papers), National Archives, Record Group 77, Washington, D.C. (hereafter cited as NA, RG 77).

7. Abert to Secretary of War Joel Poinsett, March 31, 1840, Letters Sent by the Topographical Bureau, 1829–1863 (hereafter cited as LSTB), entry 310, NA, RG 77.

8. James L. Morrison, "Educating the Civil War Generals: West Point, 1833–1861," *Military Affairs* 38 (October 1974): 108; for the Corps' preferred status, see also "Military Academy," *The North American Review* 57 (October 1843): 269–292.

9. Ellicott continued: "Scarcely two compasses will trace with the necessary precision the same lines, and for want of a general standard, by which the chains should be regulated, it is difficult to find two the same length"; see *Several Methods by Which Meridian Lines May Be Found with Ease and Accuracy* (Philadelphia: Dobson, 1796), p. 5.

10. John Love, *Geodaesia: Or the Art of Surveying and Measuring Land, Made Easy* . . . (London: J. Taylor, 1688), pp. i, 191–192.

11. Love, *Geodaesia*, p. 55. Colonists often ran straight lines along rivers, ignoring contours; see Neil M. Nugent, *Cavaliers and Pioneers: Abstracts of Virginia Land Patents and Grants, 1623–1800*, vol. 1, 1623–1626 (Richmond, Va.: Press of the Deity Print Co., 1934), p. 3; for folkways of the chain surveyor, see William Leybourn, *The Complete Surveyor: Containing the Whole Art of Surveying Land* (London: R. & W. Leybourn, 1657); for John Love's popular handbook, see Louis Karpinski, *Bibliography of Mathematical Works Published in America through 1850* (Ann Arbor: University of Michigan Press, 1940), p. 284.

12. Byrd, "History of the Dividing Line," reprinted in Louis B. Wright, *The Prose Works of William Byrd of Westover* (Cambridge: Harvard University Press, 1966), pp. 182, 189, 190, 200.

13. Darwin H. Stapleton, ed., *The Engineering Drawings of Benjamin Henry Latrobe* (New Haven: Yale University Press, 1980), p. 62; for the class hierarchy, see Sarah S. Hughes, *Surveyors and Statesmen: Land Measurers in Colonial Virginia* (Richmond: Virginia Association of Surveyors, 1979), pp. 142–143.

14. Silvio A. Bedini, *Tinkers and Thinkers: Early Men of Science* (New York: Charles Scribner's Sons, 1975), p. 160.

15. Ibid., pp. 160–161; see also I. Bernard Cohen, "How Practical Was Benjamin Franklin's Science?" *Pennsylvania Magazine of History and Biography* 6 (1945): 284–293.

16. John Clarence Webster, "Joseph Frederick Wallet Des Barres and the Atlantic Neptune," *Proceedings and Transactions of the Royal Society of Canada*, series 3 (1927), vol. 21, section 2, pp. 21–40; see also Willis Chipman, "The Life and Times of Major Samuel Holland, Surveyor-General, 1764–1801," *Ontario Historical Society Papers and Records* 21, no. 11 (1924), pp. 11–90; and Louis De Vorsey, "Hydrography: A Note on the Equipage of the Eighteenth Century Survey Vessel," *Mariner's Mirror* 58 (1972): 173–178.

17. Henry De Bernierre, *General Gage's Instructions of 22d February, 1775, to Captain Brown and Ensign D'Berniere . . . Together with an Account of Their Doings* (Boston: J. Gill, 1779), p. 311.

18. Hugh Debbeig's critique of the Holland expedition, January 15, 1764, quoted in Marshall, "British Military Engineers," p. 230; see also Clarence Edwin Carter, ed., *The Correspondence of General Thomas Gage with the Secretaries of State and with the War Office and the Treasury, 1763–1775*, 2 vols. (New Haven: Yale University Press, 1931), 1:309, 310, 347.

19. John B. Hartley, Barbara B. Patchenick, and Lawrence W. Towner, *Mapping the American Revolutionary War* (Chicago: University of Chicago Press, 1978), p. 70.

20. Philip Pittman, *The Present State of the European Settlements on the Mississippi with a Geographical Description of That River, Illustrated by Plans and Documents* (London: Printed for J. Nourse, 1770); see also William P. Cummings, *British Maps of Colonial America* (Chicago: University of Chicago Press, 1974), pp. 36–37, 57.

21. Bedini, *Tinkers and Thinkers,* pp. 355–366; see also Walter W. Ristow, "John Mitchell's Map of the British and French Dominions in North America," in Ristow, comp., *A la Carte: Selected Papers on Maps and Atlases* (Washington, D.C.: Library of Congress, 1972), pp. 103–113; and Louis de Vorsey, "Pioneer Charting of the Gulf Stream: The Contributions of Benjamin Franklin and William Gerard De Brahm," *Image Mundi* 28 (1976): 105–120.

22. Anna Margaret Quattrocchi's dissertation on "Thomas Hutchins, 1735–1789" (University of Pittsburgh, 1944) is the most complete biography; see also Frederick Charles Hicks, "Biographical Sketch of Thomas Hutchins," in Hutchins, *A Topographical Description of Virginia, Pennsylvania, Maryland, and North Carolina* (Cleveland: The Burrows Brothers Company, 1904), pp. 7–51, and Joseph G. Tregle, Jr., introduction to Hutchins, *An Historical Narrative and Topographical Description of Louisiana, and West-Florida,* facsimile reproduction of the 1784 ed. (Gainesville: University of Florida Press, 1968), pp. v–xlvii.

23. Hutchins, *A Topographical Description,* pp. 70, 80.

24. Pittman, *Present State of European Settlements,* pp. 34–36; compare with Hutchins's view of Balize Bayou in *An Historical Narrative,* pp. xlii, 33–34; see also Hutchins, "Description of a remarkable rock and cascade," *Transactions,* American Philosophical Society, 2 (1786): 50; and Gilbert Imlay, *Topographical Description of the Western Territories* (New York: Samuel Campbell, 1793), pp. 304–305.

25. Franklin to the president of Congress, March 16, 1780, in Jared Sparks, ed., *The Works of Benjamin Franklin,* 10 vols. (Boston: Tappun & Whittemore, 1844), 8:436–437.

26. U.S. Congress, Resolution, July 11, 1781, Box 3, folio 6, Papers of Thomas Hutchins, Geographer General, Historical Society of Pennsylvania, Philadelphia; for Hutchins's "treasonable correspondence," see William Bell Clark, "In Defense of Thomas Diggs," *Pennsylvania Magazine of History and Biography* 77 (1953): 400.

27. Hicks, "Biographical Sketch of Thomas Hutchins," p. 32.

28. The first three U.S. surveyor generals—Hutchins, Rufus Putnam, and Jared Mansfield—all served as engineer-officers; see William D. Pattison, *Beginnings of the American Rectangular Land Survey System, 1784–1800,* University of Chicago Department of Geography, Research Paper no. 50 (Chicago: University of Chicago Press, 1964).

29. Hutchins, *An Historical Narrative,* pp. 33–34.

30. Hutchins wrote *An Historical Narrative* as a companion to his map of West Florida, but apparently there were not enough advanced subscriptions to warrant the map's publication; see Thomas Hutchins, *Proposal for Publishing . . . a Map of . . . West Florida . . . October 15, 1781,* a broadside (Philadelphia: Printed for the author by Robert Aitken, 1781); for the eighteenth-century travelogue as a literary genre, see Percy G. Adams, *Travelers and Travel Liars, 1660–1880* (Berkeley and Los Angeles: University of California Press, 1962), p. 223.

31. The Fitch map is reprinted in Philip L. Phillips, *The Rare Map of the North-west, 1785* (Washington: W. H. Lowdermilk, 1918), p. 7.

32. Jedidiah Morse, *The American Geography* (London: Printed for John Stock-dale, 1794), p. vi; see also R. H. Brown, "The American Geographies of Jedidiah Morse," *Annals of the Association of American Geographers* 31 (1941): 145–217, and John Filson, *Discovery, Settlement, and Present State of Kentucky* (New York: Samuel Campbell, 1793), frontispiece.

33. John L. Allen "Geographical Knowledge and American Images of the Louisiana Territory," *Western Quarterly Review* 2 (April 1981): 156; see also Elliott Coués, ed., *The Expedition of Zebulon Montgomery Pike*, 2 vols. (New York: Francis P. Harper, 1895), 2:525.

34. Tregle, introduction to Hutchins, *An Historical Narrative*, p. xliv.

35. Hutchins, *An Historical Narrative*, p. 29.

36. Ibid., p. 94.

37. Ibid., p. 29.

38. Quattrocchi, "Thomas Hutchins," pp. 281–290; see also Tregle, introduction to Hutchins, *An Historical Narrative*, p. xxxiv.

39. Allen, "Geographical Knowledge and American Images," p. 163.

40. Washington to Hutchins, August 20, 1786, in Jared Sparks, ed., *The Writings of George Washington*, 32 vols. (New York: Harper & Brothers, 1847), 9:195.

41. U.S. Constitution, Article 1, Section 8.

42. James Madison on the military establishment, 1794, quoted in Theodore J. Crackel, *Mr. Jefferson's Army: Political and Social Reform of the Military Establishment, 1800–1809* (New York: New York University Press, 1987), p. 11; see also A. Hunter Dupree, *Science in the Federal Government: A History of Polities and Activities* (Baltimore: Johns Hopkins University Press, 1986), pp. 4–6.

43. Albert H. Heusser, *George Washington's Map Maker: A Biography of Robert Erskine* (New Brunswick, N.J.: Rutgers University Press, 1966), p. 165.

44. William Tatham, *The Political Economy of Inland Navigation, Irrigation, and Drainage . . . in the Remotest Interior of Great Britain and Foreign Parts* (London: Faulder, 1799), p. 1.

45. The best account is G. Melvin Herndon, *William Tatham, 1752–1819: American Versatile* (Johnson City: East Tennessee State University, 1973); for a list of his publications, see "William Tatham," *The Annual Biography and Obituary for the Year 1820* 4 (1820): 149–168.

46. Tatham, *Political Economy of Inland Navigation*, p. 306.

47. Herndon, *William Tatham*, p. 177; see also Phillips, *Inland Navigation*, p. 111.

48. Jefferson to Secretary of War Henry Dearborn, August 31, 1807, in *The Writings of Thomas Jefferson*, ed. Andrew A. Lipscomb, vol. 11 (Washington, D.C.: Thomas Jefferson Memorial Association, 1905), p. 349.

49. William Tatham, *Statement Explanatory of the Public Interest in William Tatham's Topographical and Other Employments* (Washington, D.C.: J. Grossfield, 1813), pp. 6–7.

50. Herndon, *William Tatham*, p. 238; see also Samuel C. Williams, *DAB*, s. v. "Tatham, William."

51. The story has a tragic ending. On February 22, 1819, Tatham stepped in front of a rifleman just as he fired a salute. In *DAB* Samuel Williams indicates suicide; see

also Tatham to the president of the United States, July 11, 1812, and Tatham to John Armstrong, July 2, 9, 13, 22, 30, 1814, Letters Received, Office of the Secretary of War, War Department (hereafter cited as LRSW), M-221, 1813–1863, NA, RG 107.

52. Tatham, *Statement,* p. 8.

53. "Fortifications at Charleston . . . Instructions to Paul Hyacinte Perrault [elsewhere spelled Parrault and Perreault]," H. Doc. 22, 3d Cong., 2d sess. (1794), pp. 101–103. Brief notes on John Anderson, J. J. Abert, and other topographers can be found in the biographical files of the Corps of Engineers Office of History at Fort Belvoir, Va; for Calhoun's staff reorganization, see Russell F. Weigley, *History of the United States Army* (New York: Macmillan Publishing Co., 1967), pp. 133–139.

54. James Renwick to Joseph Swift, May 13, 1816, mss 69, West Point Library, West Point, New York.

55. Karl Bernard, duke of Saxe-Weimar-Eisenach, *Travels through North America, during the years 1825 and 1826,* 2 vols. (Philadelphia: Carey, Lee, and Carey, 1828), 1:181.

56. Kennedy, *Orders from France,* p. 98.

57. Roberdeau Buchanan, *Genealogy of the Roberdeau Family* (Washington, D.C.: Joseph L. Pearson, 1876), pp. 104–122; see also *DAB,* s.v. "Roberdeau Isaac," and, Francis B. Heitman, *Historical Register and Dictionary of the U.S. Army from Its Organization, September 29, 1789 to March 2, 1903,* 2 vols. (Washington: U.S. Government Printing Office, 1903), 1:834.

58. Buchanan, *Genealogy of the Roberdeau Family,* p. 113; see also Joseph Gardner Swift, *The Memoirs of General Joseph Gardner Swift, LL.D., U.S.A., First Graduate of the United States Military Academy, West Point,* ed. H. Ellery (Worcester, Mass.: Press of F. S. Blanchard and Co., 1890), pp. 44, 82, 204, 205.

59. Isaac Roberdeau, "Mathematics and Treatise on Canals," unpublished, n.d. (c. 1796), Manuscript Division, Library of Congress, p. 128.

60. Extract from Major Roberdeau's "Report of U.S. Topographical Engineers to the Honorable Thomas Newton" (1824), p. 3, Correspondence of the Topographical Officer Stationed in Washington, D.C., 1816–1826, Entry 306 (hereafter cited as Roberdeau Papers), NA, RG 77; see also Col. Isaac Roberdeau, "Observations on the Survey of the Sea Coast of the United States" (Paper read before the Columbia Institute, December 9, 1826), miscellaneous pamphlet in the Rare Book Room of the Library of Congress (n.p., 1826?); for a reference to the Ponts et chaussées, see Abert's report of 1834 in "Topographical Engineers," H. Rept. 95, 24th Cong., 1st sess. (1836), p. 14.

61. Abert to Secretary of War Peter B. Porter, February 12, 1829, LSTB, NA, RG 77; see also Hugh Young, "A Topographical Memoir on East and West Florida With Itineraries of General Jackson's Army, 1818," introduction by Mark F. Boyd and Gerald M. Poton, *Florida Historical Quarterly* 13 (July, October 1934): 16–50, 129–164; for the wage dispute, see *ASP, MA* 2:296, 303.

62. Alexander Macomb, "Report of the operation of the engineer department during the year ending September 30, 1825," *ASP, MA* 3:140.

63. Roger L. Nichols and Patrick L. Halley, *Stephen Long and American Frontier Exploration* (Newark: University of Delaware, 1980), p. 17; for a full biography, see Richard G. Wood, *Stephen Harriman Long, 1784–1864: Army Engineer, Explorer, Inventor* (Glendale, Calif.: Arthur H. Clark Co., 1966).

64. *Memoirs of General Joseph Swift,* p. 204.

65. Long to Thomas A. Smith, September 15, 1816, LRSW Main Series, 1800–1870, NA, RG 107.

66. Stephen H. Long, "Canal to Connect the Illinois River with Lake Michigan," *American State Papers, Miscellaneous,* 2:555–556.

67. Long to George Graham, March 4, 1817, LRSW, NA, RG 107; see also Richard G. Wood, "Stephen Long's Plan for New Fort at Peoria," *Journal of the Illinois State Historical Society* 47 (Winter 1954): 417–421.

68. Long to James Monroe, March 15, 1817, LRSW, NA, RG 107.

69. Lucile M. Kane, June D. Holmquist, and Carolyn Gilman, eds., *The Northern Expeditions of Stephen H. Long: The Journals of 1817 and 1823 and Related Documents* (St. Paul: Minnesota Historical Society, 1978), pp. 70, 77.

70. Stephen H. Long, "Voyage in a Six-Oared Skiff to the Falls of Saint Anthony in 1817," *Collections of the Minnesota Historical Society* 2 (1860–1867, reprinted 1889): 9–88; for a reprint with reference notes, see Kane, Holmquist, and Gilman, *Northern Expeditions of Stephen Long,* pp. 49–110; see also Long to Smith, May 12, 1818, Buell Collection of Engineer Historical Papers, 1801–1819, Entry 292 (hereafter cited as Buell Collection; also available on Archives Microcopy No. 417), NA, RG 77.

71. Kane, Holmquist, and Gilman, *Northern Expeditions of Stephen Long,* pp. 55, 66.

72. Long to the Hon. president and members of the American Philosophical Society, March 7, 1817, and F. Beasley, et al., Report of the Committee to whom was referred the letter of Stephen H. Long, March 7, 1810; Miscellaneous Correspondence, Archives of the American Philosophical Society, Philadelphia.

73. Long to Thomas Smith, May 12 1818; reprinted in Kane, Holmquist, and Gilman, *Northern Expeditions of Stephen Long,* p. 335.

74. Kane, Holmquist, and Gilman, *Northern Expeditions of Stephen Long,* p. 344.

75. Preliminary notice to Edwin James, comp., *Account of an Expedition from Pittsburgh to the Rocky Mountains . . . under the Command of Major Stephen H. Long,* 2 vols. with an atlas (Philadelphia: H. C. Carey & I. Lea, 1823), 1:3–4; see also Long to Secretary of War Calhoun, December 24, 30, 1818, LRSW, NA, RG 107; and "Report on Military Peace Establishment," H. Doc. 36, 15th Cong., 2d sess. (1818); for the Yellowstone expedition, see Hiram M. Chittenden, *A History of the American Fur Trade of the Far West,* 2 vols. (Stanford: Stanford University Press, 1954), 2:562–587; see also Goetzmann, *Army Exploration in the West,* pp. 39–45.

76. Niles *Weekly Register,* n.s., 4 (July 24, 1819): 368.

77. Long reports on the steamboat in letters to Secretary of War Calhoun, January 29 to October 28, 1819; LRSW, NA, RG 107; see also Roger L. Nichols, "Army Contributions to River Transportation, 1818–1825," *Military Affairs* 33 (April 1969): 242–249.

78. James, *Account of an Expedition* 2:361.

79. "Account of an Expedition," *North American Review,* n.s., 7 (1823): 242.

80. Carl I. Wheat, *Mapping the Transmississippi West, 1540–1861,* 5 vols. (San Francisco: Institute of Historical Cartography, 1957–1963), 2:78–81; Herman R. Friis maintains that Long, not Swift, was the primary cartographer in "Stephen Long's Unpublished Manuscript Map of the United States Compiled in 1820–1822," *California Geographer* 8 (1967): 75–87.

81. James, *Account of an Expedition* 2:45; see also John R. Bell, *The Journal of Captain John R. Bell,* The Far West and the Rockies Historical Series, ed. Harlin M. Fuller and Leroy R. Hafen, 15 vols. (Glendale, Calif.: Arthur H. Clark Co., 1957), p. 178.

82. Henry Schenck Tanner, *A New American Atlas* (Philadelphia: H. S. Tanner, 1823). A copy of the original by Swift and Long can be found in the Cartographic Division of the National Archives, Alexandria, Virginia.

83. James Fenimore Cooper, *The Prairie: A Tale,* 2 vols. (Philadelphia: Carey, Lea & Carey, 1827), 1:16; see also James, *Account of an Expedition* 1:454; Orm Overland makes this comparison in "James Fenimore Cooper's *The Prairie:* The Making and Meaning of the American Classic" (Ph.D. diss., Yale University, 1969), p. 123.

84. Lyman Frank Baum, *The Wonderful Wizard of Oz* (Chicago: G. M. Hill, 1900), p. 12. Parkman borrowed the image from James in describing the basin of the Platte as "a barren trackless waste"; see *The California and Oregon Trail, Being Sketches of Prairie and Mountain Life* (New York: A. L. Burt Co., 1966), p. 56.

85. James, *Account of an Expedition* 2:361; for the desert-as-barrier concept, see Richard H. Dillon, "Stephen Long's Great American Desert," *Proceedings of the American Philosophical Society* 3 (April 14, 1967): 93–108; see also Henry Nash Smith, *Virgin Land* (Cambridge: Harvard University Press, 1950), pp. 174–183.

86. William Hypolitus Keating, *Narrative of an Expedition to the Source of the St. Peter's River . . . under the Command of Stephen H. Long,* 2 vols. (Philadelphia: H. C. Carey & I. Lea, 1824), 1:232.

87. Jared Sparks, "Major Long's Second Expedition," *North American Review* 21 (July 1825): 185. Explorer Henry Schoolcraft criticized Long's expedition; see "La Decouverte des Sources du Mississippi," *North American Review* 27 (July 1828): 178–189.

88. W. Eugene Hollon, *The Great American Desert Then and Now* (Oxford: Oxford University Press, 1966), pp. 64–65.

89. Keating, *Narrative of an Expedition* 2:232.

90. Keating, *Narrative of an Expedition* 1:231.

91. G. Malcolm Lewis, "Regional Ideas and Reality in the Cis-Rocky Mountain West," *Transactions,* Institute of British Geographers, 38 (June 1966): 138.

92. Goetzmann, *Exploration and Empire,* p. xii; see also William H. Goetzmann, *New Lands, New Men: America and the Second Great Age of Discovery* (New York: Viking, 1986), pp. 120–126.

93. Reuben Gold Thwaites, ed., *Original Journals of the Lewis and Clark Expedition, 1804–1806,* 8 vols. (New York: Dodd, Mead, & Co., 1904–1905), 7:310.

94. Keating, *Narrative of an Expedition* 1:242, 2:191; see also Wyndham D. Miles, "A Versatile Explorer: A Sketch of William H. Keating," *Minnesota History* 36 (December 1959): 294–299.

95. Totten to John Calhoun, February 8, 1824, Totten Papers, 1:17–23, NA, RG 77.

96. Long to Isaac Roberdeau, May 31, 1824, Roberdeau Papers, NA, RG 77. Long apologized for his indignation in a letter to Roberdeau two days later (July 2, 1824); for Macomb's attempt to make Long pay for unauthorized expenses from the 1820 expedition, see John Livingston, "Col. Stephen H. Long, of the U.S. Army," in *Portraits of Eminent Americans Now Living,* 4 vols. (New York: Cornish, Lamport & Co., 1854) 4:489.

97. Kane, Holmquist, and Gilman, *The Northern Expeditions of Stephen Long,* p. 43.

98. Report of the committee on Major Long's substitute for locks in canals, March 4, 1825; Miscellaneous Communications, American Philosophical Society Library, Philadelphia; for Long's barometer experiments see Nichols and Halley, *Stephen Long and Frontier Exploration,* pp. 178–179.

99. When McRee declined the appointment, the position did indeed go to Crozet; see Long to Roberdeau, February 3, 1823, Roberdeau Papers, NA, RG 77.

100. Guillaume Tell Poussin, *The United States: Its Power and Progress,* trans. Edmund L. DuBarry (Philadelphia: Lippincott, Grambo, and Co., 1851), p. 400; see also Long to Roberdeau, June 2, 1824, Roberdeau Papers, NA, RG 77.

101. The figures are from 1826; see Roberdeau's semiannual reports, documents 6R, 19R, 46R, and 54R, series E (1826–1837), Letters Received, Office of the Chief of Engineers, entry 18, NA, RG 77.

102. John Kirkland Wright, *Human Nature in Geography* (Cambridge: Harvard University Press, 1966), p. 224.

103. Nichols and Halley, *Stephen Long and Frontier Exploration,* p. 36.

104. Wright makes this point in *Human Nature in Geography,* p. 33.

3. The West Point Connection

1. Captain Hughes, a future congressman, spent four years at West Point but did not graduate; see George W. Hughes, "Survey of the Ohio River," H. Doc. 50, 27th Cong., 3d sess. (1843), p. 10.

2. Dixon Ryan Fox, "Civilization in Transit," *American Historical Review* 32 (July 1927): 753–768; for West Point as our Athens and Sparta, see Joseph Ellis and Robert Moore, *School for Soldiers: West Point and the Profession of Arms* (New York: Oxford University Press, 1974), p. 32.

3. U.S., *Statutes at Large* 1:366–367; see also U.S. Congress, *Annual of Congress,* 1st Cong., 1st sess. (1790), pp. 2089–2091.

4. Timothy Pickering quoted in Arthur Pearson Wade, "Artillerists and Engineers: The Beginnings of American Seacoast Fortifications, 1794–1815" (Ph.D. diss., Kansas State University, 1977), p. 77.

5. U.S. Congress, *ASP, MA* 1:129.

6. *Works of Alexander Hamilton,* ed. Henry Cabot Lodge, 9 vols. (New York: G. P. Putnam's Sons, 1903), 7:178–179; see also Richard H. Kohn, *Eagle and Sword: The Federalists and the Creation of the Military Establishment in America, 1783–1802* (New York: Free Press, 1975), pp. 17–73.

7. U.S., *Statutes at Large* 2:137.

8. U.S. Congress, *ASP, MA* 1:156; see also James L. Morrison, Jr., *The Best School in the World* (Kent, Ohio: Kent State University Press, 1986), pp. 141–147.

9. Theodore Crackel says Jefferson used the 1802 reorganization act to "sever Federalist control of the army's internal administration"; see *Mr. Jefferson's Army,* p. 45, passim; see also Donald R. Hickey, "Federalist Defense Policy in the Age of Jefferson, 1801–1812," *Military Affairs* 45 (April 1981): 63–70; for West Point as a science academy, see Stephen Ambrose, *Duty, Honor, Country: A History of West Point* (Baltimore: Johns Hopkins University Press, 1966), p. 18.

10. Mildred E. Lombard, *DAB* s.v. "Williams, Jonathan"; see also, Sidney Forman, "The United States Military Philosophical Society, 1802–1813," *William and Mary Quarterly,* 3d series, 2–3 (July 1945): 278–279; for Williams's resemblance to

Franklin, see *Memoirs of General Joseph Swift,* p. 23; see also Dorothy J. Zuersher, "Benjamin Franklin, Jonathan Williams, and the United States Military Academy" (Ph.D. diss., University of North Carolina at Greensboro, 1974), p. 60; for the barometer experiments, see Silvio A. Bedini, *Thomas Jefferson: Statesman of Science* (New York: Macmillian Publishing Co., 1990), pp. 259–260.

11. Jonathan Williams, Address on laying first foundation stone of Ft. Tompkins, May 26, 1814, Jonathan Williams Manuscripts.

12. Jonathan Williams, "Address," January 30, 1808, in *Extracts from the Minutes of the United States Military Philosophical Society,* Microfilm Reading Room, Library of Congress, Washington, D.C.; see also Jonathan Williams, trans., *The Elements of Fortification,* 2d ed. (Philadelphia: C. P. Wayne, 1801).

13. Edgar Denton III, "The Formative Years of the United States Military Academy, 1775–1833" (Ph.D. diss., Syracuse University, 1964), p. 31.

14. Williams to Decius Wadsworth, August 13, 1802, Jonathan Williams Manuscripts; Sen. Abraham Baldwin, a Georgia Republican, said the Corps of Engineers would be "a little selected band . . . like the corps de Genie In Europe", quoted in Charles Hendricks, "The Early Years of the U.S. Military Academy" (Paper read at the American Military Institute meeting in Lexington, Kentucky [1989]), p. 4.

15. Hamlin, *Benjamin Henry Latrobe,* p. 386; for Tousard, see Norman B. Wilkinson, "The Forgotten 'Founder' of West Point," *Journal of Military History* 24 (Winter 1960–61): 177–188; for ordnance standardization, see Merritt Roe Smith, "Military Entrepreneurship," in Otto Mayr and Robert C. Post, eds., *Yankee Enterprise: The Rise of the American System of Manufactures* (Washington, D.C.: Smithsonian Institution, 1981), pp. 66–67.

16. Samuel F. Tilman, "The Academic History of the Military Academy, 1802–1902," *The Centennial History of the United States Military Academy at West Point, New York,* 2 vols. (Washington, D.C.: U.S. Government Printing Office, 1904), 1:275.

17. Charlotte W. Dudley, "Jared Mansfield: United States Surveyor General," *Ohio History* 85 (Summer 1976): 231–246; see also *DAB* s.v. "Mansfield, Jared"; Hunter and Dooley, *Claudius Crozet,* p. 23, and Charles Davies, *Elements of Surveying* (New York: J. & J. Harper, 1830).

18. M. L. Welch, "Early West Point: French Teachers and Influences," *The American Society Legion of Honor Magazine* 26 (Spring 1955): 32; for Hassler, see Martha Coleman Bray, *Joseph Nicollet and His Map* (Philadelphia: The American Philosophical Society, 1980), pp. 51–53; see also Florian Cajori, *The Chequered Career of Ferdinand Rudolph Hassler* (Boston: Christopher Publishing House, 1929).

19. Andrew Ellicott to brother Joseph, October 17, 1813, in Catherine Mathews, *Andrew Ellicott: His Life and Letters* (New York: Grafton Press, 1908), pp. 230, 243–244.

20. Charles Hutton, *A Course of Mathematics,* 2 vols. (New York: Samuel Campbell, 1822) 2:59.

21. "Inventory of Mathematical and Philosophical Instruments and Inventory of Books, Maps, and Charts," July 11, 1803, Jonathan Williams Manuscripts, Lilly Library, Indiana University, Bloomington; see also Barron to Jonathan Williams, March 11, 1803, Jonathan Williams Manuscripts.

22. Hutton, *Course of Mathematics,* 2:59, 255.

23. Hutton's text was used at the British Woolwich academy, and Enfield's was adopted by Harvard; for a cadet's view of the early texts, see Edward Deering Mansfield, *Personal Memories* (Cincinnati: Robert Clarke Co., 1819), p. 65; see also "Remarks on Dr. Enfield's *Institutes of National Philosophy,*" *American Journal of Science* 3 (1821): 126.

24. Jonathan Williams, "A Thermometrical Journal of the temperature of the atmosphere and Sea . . . ," *Tramsactions,* American Philosophical Society, 3 (1793): 194–202; quoted in Konvitz, *Cartography in France,* p. 66.

25. Jonathan Williams, *Thermometrical Navigation; Being a Series of Experiments and Observations, Tending to Prove, that by Ascertaining the Relative Heat of the Sea-Water from Time to Time, the Passage of a Ship through the Gulf Stream and from Deep Water into Soundings, may be Discovered* . . . (Philadelphia: R. Aitken, 1799), pp. 41–42.

26. Jonathan Williams, "Original Nautical Thermometrical Journal, October 6, 1806," *Extracts from the Minutes of the United States Military Philosophical Society;* see also Benjamin Franklin, "Letter Containing Sundry Maritime Observations," *Transactions,* American Philosophical Society, 2 (1786): 50.

27. Crackel, *Mr. Jefferson's Army,* pp. 59–60.

28. Williams to Thomas Jefferson, December 12, 1802, United States Military Philosophical Society, Minutes and Records, Membership Lists, Correspondence and Papers, New York Historical Society, New York. Although a fire at West Point may have destroyed some of the minutes in 1838, the New York Historical Society has four volumes of records. This study relied on photocopies in the U.S. Army Corps of Engineers Office of History and the microfilmed *Extracts* of the society's papers.

29. "Circular," December 20, 1806, *Extracts from the Minutes of the United States Military Philosophical Society;* see also Arthur P. Wade, "A Military Offspring of the American Philosophical Society," *Military Affairs* 38 (September 1974): 103–107.

30. Forman, "United States Military Philosophical Society," p. 280.

31. The full record of the October 6, 1806, meeting is in the handwritten U.S. Military Philosophical Society's Minutes and Records (see note 28).

32. Historian Sidney Forman says William Barron read this paper, but it was clearly based on Jonathan Williams's 1801 study of the Ohio; see Dwight L. Smith, ed., "The Ohio River in 1801: Letters of Jonathan Williams, Junior," *Filson Club Historical Quarterly* 27 (July 1953): 199–222.

33. David A. Clary, *Fortress America: The Corps of Engineers, Hampton Roads, and United States Coastal Defense* (Charlottesville: University Press of Virginia, 1990), pp. 29–32; see also Wade, "Artillerists and Engineers," pp. 198–199, 208, and *Annals of Congress* 15:1227.

34. Robert S. Browning III, *Two If by Sea: The Development of American Coastal Defense Policy* (Westport, Conn.: Greenwood Press, 1983), pp. 17–81; see also A Plan of the entrance into Winyawh Bay, at Georgetown, folio 1808, Jonathan Williams Manuscripts, and for the future chief engineers, see George W. Cullum, *Biographical Register of the Officers and Graduates of the United States Military Academy at West Point, New York,* 3 vols. (New York: James Miller, 1868), 1: passim. Macomb, although never a cadet, studied under Williams as a lieutenant at West Point from 1801 to 1805; see, *DAB,* s.v. "Macomb, Alexander."

35. George W. Cullum, *Campaigns of the War of 1812–15, against Great Britain Sketched and Criticized; with Brief Bibliographies of the American Engineers* (New York: James Miller, publisher, 1879), p. 41.

36. Williams to Callander Irvine, July 19, 1806; quoted in Charles Walker, "Academic Preparation" (a draft prepared for the Office of History, U.S. Army Corps of Engineers [typewritten, n.d.]), p. 26.

37. Wade, "Artillerists and Engineers," p. 282; see also Forman, "United States Military Philosophical Society," p. 284.

38. Henry Adams, *History of the United States of America during the Administrations of Thomas Jefferson and James Madison,* 9 vols. (New York: Charles Scribner's Sons, 1889–1891), 9:236.

39. Ambrose, *Duty, Honor, Country,* p. 43.

40. Cullum, *Campaigns of the War of 1812,* pp. 165–168.

41. A West Point library inventory of 1822 listed only 942 titles. Perhaps some of the books never made it to West Point, or Thayer may have purchased many copies of the same title for classroom use; see *Catalogue of Books, in the Library of the Military Academy* (Newburgh, N.Y.: Ward H. Gazlay, 1822); see also Sylvanus Thayer, Books from Europe and Equipment at West Point, n.d., Letters and Papers Received (irregular series), 1789–1831, entry 20, NA, RG 77.

42. Richard E. Dupuy, *Where They Have Trod: The West Point Tradition in American Life* (New York: Frederick A. Stokes Co., 1940), pp. 156–159; for the West Point debt to the Ecole polytechnique see William H. C. Bartlett's summary of the conversation with Thayer, recorded years later by George Cullum (September 26, 1877) and included chronologically in Cindy Adams, ed., "The West Point Thayer Papers, 1808–1872," West Point, New York, Association of Graduates, 1965 (typewritten); see also William James Kershner, "Sylvanus Thayer: A Biography" (Ph.D diss., West Virginia University, 1976), pp. 73–99, and Peter Michael Molloy, "Technical Education and the Young Republic: West Point as America's Ecole Polytechnique, 1802–1833" (Ph. D. diss., Brown University, 1971), p. 372, passim.

43. Letter to William McRee and Sylvanus Thayer, January 4, 1816, Buell Collection, entry 292, NA, RG 77.

44. For the Calhoun-Hamilton comparison, see Weigley, *History of the United States Army,* pp. 133–143; see also Richard K. Cralle, ed., *The Works of John C. Calhoun,* 6 vols. (New York: D. Appleton and Co., 1851–1856) 5:54–57.

45. Mansfield, "United States Military Academy," p. 39; see also Robert H. Hall, "Early Discipline at the United States Military Academy," *Journal of the Military Service Institute of the United States* 2 (1882): 452–458, and Hunter and Dooley, *Claudius Crozet,* p. 28.

46. *DAB,* s.v. "Crozet, Claude."

47. The description of Crozet is taken from John H. B. Latrobe's memoir; quoted in Colonel William Couper, *Claudius Crozet: Soldier-Scholar-Educator-Engineer, 1789–1864,* Southern Sketches no. 8 (Charlottesville, Va: Historical Publishing Co., 1936), p. 24; see also Hunter and Dooley, *Claudius Crozet,* p. 17.

48. Couper, *Claudius Crozet,* p. 29.

49. Molloy, "Technical Education in the Young Republic," p. 371.

50. Claudius Crozet, *Treatise on Descriptive Geometry* (New York: A. T. Goodrich & Co., 1821); see also *A History of Mathematics Education in the United States and*

Canada (Washington, D.C.: National Council of Teachers of Mathematics, 1970), pp. 28–29.

51. Sganzin, *Elementary Course of Civil Engineering*, p. 192, passim; see also *La France Litteraire* (Paris: Chez Firmin Didot Frepes, 1838), p. 115, and Ralph R. Shaw, "Engineering Books Available in America Prior to 1830," *Bulletin of the New York Public Library* 37 (1933): 546–547.

52. Sganzin, *Elementary Course of Civil Engineering*, from the translator's preface.

53. Evald Rink lists Sganzin's 1827 Boston edition as the first American publication on civil engineering; see Rink, *Technical Americana: A Checklist of Technical Publications Printed before 1831* (Millwood, N.Y.: Kraus International Publications, 1981), p. 199.

54. Amos Eaton, *Art without Science; or, Mensuration, Surveying and Engineering Divested of the Speculative principles and Technical Languages of Mathematics*, 2d ed. (Albany: Webster & Skinner, 1830).

55. Rink's *Technical Americana* has no record of Eaton's first edition, which was presumably printed locally as a complement to his lectures at Rensselaer. The tallies of graduates were compiled from Cullum's *Biographical Register*.

56. "Academy at West Point," *American Quarterly Review* 32 (December 1834): 369.

57. Lawrence P. Grayson, "A Brief History of Engineering Education in the United States," *Engineering Education* 68 (December 1977): 246–264.

58. Thomas Everett Griess, "Dennis Hart Mahan: West Point Professor and Advocate of Military Professionalism, 1830–1871" (Ph.D. diss., Duke University, 1969), p. 84.

59. Ibid., p. 122; see also Frederick A. Mahan, "Professor Dennis Hart Mahan," *Professional Memoirs, Corps of Engineers, U.S. Army and Engineer Department At Large* 9 (1917): 154–184, and Cullum, *Biographical Register*, pp. 319–325.

60. West Pointers Henry Halleck, John G. Barnard, Andrew Humphreys, Sylvanus Thayer (in 1843), and others took furloughs in France to see doctors and convalesce; see Dale E. Floyd, "U.S. Army Officers in Europe, 1815–1861," *Proceedings of The Citadel Conference on War and Diplomacy*, ed. David H. White and John W. Gordon (Charleston, S.C.: The Citadel, 1977), p. 28.

61. Griess, "Dennis Hart Mahan," p. 122.

62. Lafayette opened doors for Mahan in Paris; see Russell M. Jones, "The Flowering of a Legend: Lafayette and the Americans, 1825–1834," *French Historical Studies* 4 (1966): 384–410; see also Mahan to Chief Engineer Alexander Macomb, June 20, July 25, October 9, 1827; Mahan to Secretary of War James Barbour, March 25, 1828, Textual Records of the Office of the Chief of Engineers, Letters Received, 1826–1866, entry 18, N.A. RG 77

63. "Academy at West Point," p. 370.

64. Dennis Hart Mahan, *An Elementary Course of Civil Engineering, for Use of the Cadets of the United States Military Academy* (New York: Wiley and Putnam, 1837), p. v.

65. Dennis Hart Mahan, *A Treatise on Civil Engineering*, rev. ed., with a chapter on river improvement by Capt. F. A. Mahan (New York: Wiley & Sons, 1902).

66. Griess, "Dennis Hart Mahan," p. 176.

67. F. A. Mahan, "Professor Dennis Hart Mahan," pp. 74–75.

68. Dupuy, *Where They Have Trod,* p. 199.

69. Calhoun, *American Civil Engineer,* pp. 210–217; see also Charles S. Storrow, *A Treatise on Water-Works for Conveying and Distributing Supplies of Water* (Boston: Hilliard, Grey, 1835).

70. Dennis Hart Mahan, *An Elementary Course of Civil Engineering,* 5th ed. (New York: John Wiley, 1851), pp. 317–318.

71. Dennis Hart Mahan, *A Complete Treatise on Field Fortification* (New York: Wiley & Long, 1836); see also Dennis Hart Mahan, *Industrial Drawing: Comprising the Description and Use of Drawing Instruments, the Construction of Plane Figures, the Projections and Sections of Geometrical Solids* (New York: J. Wiley, 1852).

72. D. H. Mahan, *Elementary Course of Civil Engineering,* p. vi.

73. For West Point attrition, see Cullum, *Biographical Register;* see also reports on the Military Academy in *ASP, MA* 2:51–54, 349, 350, 633, and Forman, "First School of Engineering," p. 112.

74. Mansfield, "United States Military Academy," pp. 28–29, 44; the tally excludes those who resigned before graduation or died soon after; Swift, McNeill, and Haupt can be found in the *Dictionary of American Biography;* for West Point as a source of astronomers, see Marc Rothenberg, "The Educational Background of American Astronomers, 1825–1875" (Ph.D. diss., Bryn Mawr College, 1974), pp. 10–29.

75. Stephen H. Long and William G. McNeill, *Narrative of the Proceedings of the Board of Engineers of the Baltimore and Ohio Rail Road Company . . .* (Baltimore: Bailey & Francis, 1830), pp. 140, 145.

76. John J. Abert, "Report from the Topographical Engineer, November 9, 1832," *ASP, MA* 5:64; see also O'Connell, "The Corps and Modern Management," pp. 98–103, and Hill, *Roads, Rails, and Waterways,* pp. 210–214, passim.

77. Hassler criticized army engineering in Roberdeau, "Survey of the Sea Coast of the United States," p. 21.

78. Long and McNeill, *Proceedings,* p. 159.

79. Hayden, quoted in an unpublished draft of Frank N. Schubert, "Troublesome Partnership: Gouverneur K. Warren and Ferdinand V. Hayden on the Northern Plains in 1856 and 1857," *Earth Science History* 3 (1984): 143–148; see also Charles Ellet, Jr., "Report on the Overflows of the Delta of the Mississippi," S. Exec. Doc. 20, 32d Cong., 1st sess. (1852), pp. 13–106.

80. *Army and Navy Chronicle* 10 (November 26, 1840): 346; quoted in Floyd, "U.S. Army Officers in Europe," p. 27.

81. Tallies of excursions are based on a slightly modified list generously provided by Corps historian Dale Floyd. When an officer traveled to England and France, as was the case with Thayer and Mahan, that excursion is grouped by the official or primary destination; for a breakdown of these excursions, see Todd Shallat, "Building Waterways, 1802–1861: Science and the United States Army in Early Public Works," *Technology and Culture* 31 (January 1990): 39.

82. Here the literature on hydraulic construction excludes congressional documents, river exploration, and field investigations such as the topographical corps' 1861 Mississippi Delta survey; see Robert H. Hall, comp., *Catalogue of the Library U.S. Military Academy, West Point, N.Y.* (Newburgh, N.Y.: Charles Jannicky, 1876), pp. 196–209; see also Edward S. Holden and W. L. Ostrander, "A Tentative List of Text-Books Used in the United States Military Academy at West Point from 1802 to 1902,"

Centennial of the United States Military Academy at West Point, New York 2:439–466.

83. Simon Bernard and Joseph Totten, "A Report of the Board of Engineers on the Ohio and Mississippi Rivers . . . Made in the Year 1821," H. Doc. 35, 17th Cong., 2d sess. (1823), p. 9.

84. Johnson, *Falls City Engineers,* p. 286.

85. Virginian Thomas Moore had used the term "wing dam" to describe his brush dike in the Potomac in *Ship Canal to Georgetown* (n.p., 1811), p. 8; see also Stephen H. Long, "Measures Which Have Been Taken to Improve Sand Bars of the Ohio River," H. Exec. Doc. 145, 19th Cong., 1st sess. (1826); for Long's thoughts on wing dams and river improvement, see Livingston, "Col. Stephen H. Long, of the U.S. Army," 4:490; see also Long to Robert Patterson, April 2, 1824, Miscellaneous Communications, Archives of the American Philosophical Society.

86. Long to Chief Engineer Alexander Macomb, July 17, 1824, Textual Records of the Office of the Chief of Engineers, Letters Received, 1819–1825, entry 14, NA, RG 77.

87. Hartman Bache, "Report of the Engineer Appointed to Make a Survey of the River Thames in Connecticut," H. Doc. 125, 21st Cong., 1st sess. (1930), p. 3; see also John J. Abert, "Survey of the Kennebec River," H. Doc. 108, 19th Cong., 2d sess. (1827).

88. Quincy A. Gillmore, *Practical Treatise on Limes, Hydraulic Cements, and Mortars* (New York: D. Van Nostrand, 1863); see also Gillmore, *Practical Treatise on Coignet-Béton and Other Artificial Stone* (New York: D. Van Nostrand, 1871); Joseph G. Totten, *Essay on Hydraulic and Common Mortars and on Lime-Burning, Translated from the French . . . with Brief Observations on Common Mortars, Hydraulic Mortars, and Concretes* (Philadelphia: Franklin Institute, 1838); and "Who Invented Portland Cement?" *Engineering News* 65 (April 1911): 515.

89. Sylvanus Thayer, *A Special Report on the Sea Wall . . . for the Preservation of Rams Head at the Northwest End of Lovell's Island . . .* (Washington, D.C.: Wm. Q. Force, 1844); see also Jas. St. C. Morton, *Memoir of the Life and Services of Capt. and Brevet Major John Sanders, of the Corps of Engineers, U.S. Army* (Pittsburgh: W. S. Haven, 1861), p. 45, Johnson, *Falls City Engineers,* pp. 144–145, and Gillmore, *Practical Treatise on Coignet-Béton,* pp. 49–53.

90. William Alexander Brooks, *Treatise on the Improvement of the Navigation of Rivers* (London: John Weale, 1841), pp. 6–26; see also D. H. Mahan, *Elementary Course of Civil Engineering,* 1851 ed., pp. 351–352; John G. Barnard, "The Dynamic Theory of Tides," *American Journal of Science and Arts,* 2d series, vol. 27 (May 1859), pp. 349–58, Thomas F. Hahn, *Towpath Guide to the Chesapeake & Ohio Canal,* rev. ed. (Shepherdstown, W. Va.: American Canal and Transportation Center, 1974), pp. 7–9; and Frank E. Snyder and Brian H. Guss, *The District: A History of the Philadelphia District, U.S. Army Corps of Engineers, 1866–1971* (Philadelphia: U.S. Army Engineer District, 1974), pp. 5–14.

91. For France as a conduit for British innovation, see Simon Bernard, "Essay on the Improvement of the Artillery and Ingineer [*sic*] Department of Great Britain," March 31, 1819, G-52, Letters and Papers Received (irregular series), 1789–1831, entry 20, NA, RG 77; for American neglect of Prussian reform, see Weigley, *History of the United States Army,* p. 149; see also Smith, "Military Entrepreneurship," pp. 84–85;

and Robert Bruce, *The Launching of Modern American Science, 1848–1876* (New York: Alfred A. Knopf, 1987), p. 158.

92. Review of four math textbooks, *North American Review* 13 (October 1821): 364.

93. Thomas B. Reed, *An Address . . . . to the Cadets at West Point, June 20, 1827* (New York: J. Seymour, 1827), pp. 15–16.

94. Gene D. Lewis, *Charles Ellet, Jr.: The Engineer as Individualist* (Urbana: University of Illinois, 1968), p. 15; another Paris-educated engineer was William M. Gillespie, an author, translator, and professor at Union College, 1845–1868; see Thomas N. Bonner, "The Beginnings of Engineering Education in the United States: The Curious Role of Eliphalet Nott," *New York History* 69 (January 1988): 35–54.

95. Charles Ellet, Jr., *The Mississippi and Ohio Rivers* (Philadelphia: Lippincott Grambo and Company, 1853); for French influence, see Ellet, *A Popular Notice on Wire Suspension Bridges* (Richmond, Va.: P. D. Bernard, 1839), pp. 3–8; see also Donald Sayenga, *Ellet and Roebling* (York, Pa.: American Canal and Transportation Center, 1983), pp. 11–13.

96. Laommi Baldwin II to Charles Storrow, June 7, 1830, Baldwin Papers, William L. Clements Library, University of Michigan, Ann Arbor.

97. D. H. Mahan, *Elementary Course of Civil Engineering*, 1851 ed., p. 346.

98. Bigelow learned to mix concrete as a second lieutenant under Totten at Fort Adams and also as Thayer's assistant at Boston harbor; see Cullum, *Biographical Register* 1:463; see also Peter M. Molloy, "Nineteenth-Century Hydropower: Design and Construction of Lawrence Dam, 1845–1848," *Winterthur Portfolio* 15 (1980): 315–343, and Neal Fitzsimons, "Charles S. Storrow and the Transition in American Hydraulics," *Civil Engineering* 38 (December 1968): 81–82.

99. Hughes, "Survey of the Ohio River," p. 8.

100. Edwin T. Layton, Jr., "Scientific Technology, 1845–1900: The Hydraulic Turbine and the Origins of American Industrial Research," *Technology and Culture* 20 (January 1979): 64–89; see also Andrew A. Humphreys and Henry L. Abbot, *Report upon the Physics and Hydraulics of the Mississippi River*, 2d ed. (Washington, D.C., U.S. Government Printing Office, 1867), and *DAB*, s.v. "Graham, John Duncan"; for Alexander Bache's French contacts, see Henry Blumenthal, *American and French Culture, 1800–1900: Interchange in Art, Science, Literature, and Society* (Baton Rouge: Louisiana State University Press, 1975), p. 392.

101. "Light-House Construction and Illumination," *Putnam's Monthly* 8 (August 1856): 213; for the lighthouse as a romantic symbol, see Sen. George J. Mitchell's foreword to F. Ross Holland, Jr., *Great American Lighthouses* (Washington, D.C.: Preservation Press, 1989), p. 9.

102. George Weiss, *The Lighthouse Service: Its History, Activities and Organization* (Baltimore: Johns Hopkins University Press, 1926), pp. 6–7; see also Francis Ross Holland, Jr., *America's Lighthouses: Their Illustrated History since 1716* (Brattleboro, Vt.: Stephen Greene Press, 1972), p. 32.

103. Joseph Totten, "Annual Report of the Chief Engineer, 1836," *ASP, MA* 6:857–859; see also Arnold Burges Johnson, *The Modern Light-House Service* (Washington, D.C.: U.S. Government Printing Office, 1890), pp. 16–17, and George R. Putnam, *Lighthouses and Lightships of the United States* (Boston: Houghton Mifflin Co., 1933), pp. 43–45.

104. Some historians call Brandywine Shoal (1850) the first American screw-pile, perhaps because Black Rock Ledge Beacon was a renovation job; see U.S. Light-House Establishment, *Compilation of Public Documents and Extracts from Reports and Papers Relating to Light-Houses, Light-Vessels, and Illuminating Apparatus . . .* (Washington, D.C.: U.S. Government Printing Office, 1871), p. 531; see also J. J. Abert, "Report of the Topographical Bureau," H. Doc. 2, 27th Cong., 3d sess. (1842), p. 296; J. G. Coad, "The Building of Commissioner's House, Bermuda Dockyard," *Post-Medieval Archaeology* 17 (1983): 163–176, and William H. Emory, "Screw Pile Pier," *Journal of the Franklin Institute* 8 (July 1944): 1–4.

105. Biographical Files, s.v. "Bache, Hartman," U.S. Army Corps of Engineers History Office; see also Freeman Cleaves, *Meade of Gettysburg* (Dayton, Ohio: Press of Morningside Bookshop, 1980), pp. 18, 45–47, "Early Steel Lighthouses," *Military Engineer,* 41 (July–August 1949), pp. 308–309; Holland, *America's Lighthouses,* p. 204, and Larson, *Those Army Engineers,* pp. 90–91.

106. Holland, *America's Lighthouses,* p. 100; see also Molloy, "Nineteenth-Century Hydropower," p. 322; *DAB,* s.v. "Smith, William Sooy," Charles E. Peterson, "Inventing the I-Beam: Richard Turner, Cooper & Hewitt and Others," Association for Preservation Technology, *Bulletin* 12 (1980) part 4, p. 25, and Donald J. Lehman, *Lucky Landmark: A Study of a Design and Its Survival* (Washington, D.C.: General Services Administration, 1973), pp. 13–15; for Captain Delafield's remarkable cast-iron bridge, see Daniel L. Schodek, *Landmarks in American Civil Engineering,* (Cambridge: MIT Press, 1988) pp. 79–81.

107. Holland, *America's Lighthouses,* pp. 20–21.

108. Ibid., p. 23; see also U.S. Light-House Establishment, *Compilation of Public Documents and Extracts from Reports and Papers Relating to Light-Houses,* pp. 262–263.

109. Edward Rowe Snow, *The Lighthouses of New England, 1716–1973* (New York: Dodd, Mead, & Co., 1973), pp. 192–224; see also Carl M. Guelzo, "Failure on Minot's Ledge," *Tradition* 2 (January 1959): 51–58, Schodek, *Landmarks in American Civil Engineering,* pp. 342–344, and Cullum, *Biographical Register* 1:195–196.

110. Snow, *Lighthouses of New England,* p. 206.

111. John G. Barnard, "Eulogy on the Late Joseph G. Totten," *Occasional Papers, Engineer School,* no. 16 (Washington, D.C: Press of the Engineers School, 1904), pp. 156–158; for Totten's politicking, see Kershner, *Sylvanus Thayer,* p. 298; see also *DAB* s.v. "Totten, Joseph Gilbert."

112. Schodek, *Landmarks in American Civil Engineering,* p. 344; see also Barton S. Alexander, "Minot's Ledge Lighthouse," *Transactions,* American Society of Civil Engineers, 8 (April 1879): 83–94.

113. Merritt Roe Smith, *Harpers Ferry Armory and the New Technology: The Challenge of Change* (Ithaca, N.Y.: Cornell University Press, 1977).

114. Edwin T. Layton, Jr., "Mirror-Image Twins: The Communities of Science and Technology in 19th-Century America," *Technology and Culture* 12 (October 1971): 562–580; see also Edwin T. Layton, Jr., "American Ideologies of Science and Engineering," *Technology and Culture* 17 (October 1976): 688–701; and Layton, "Scientific Technology, 1845–1900," pp. 77, 88.

115. Daniel J. Boorstin, *The Republic of Technology: Reflections on Our Future Community* (New York: Harper & Row, 1978), p. 3; see also Henry S. Commager, *The American Mind* (New Haven: Yale University Press, 1955), pp. 7–12.

116. William Gene Rothstein describes the functionalist model in "Engineers: Case and Theory in the Sociology of Professions" (Ph.D. diss., Cornell University, 1965), pp. 25–28; for an alternative model, see Rue Bucher and Anselm Strauss, "Professions in Process," *American Journal of Sociology* 66 (January 1961): 325–344.

117. Carlo M. Cipola, "The Diffusion of Innovations in Early Modern Europe," *Comparative Studies in Society and History* 14 (January 1972): 48.

118. Samuel M. Mull's historical report on the West Point drawing department in "Report of the Secretary of War," H. Doc. 2, 54th Cong., 2d sess. (1896) 1:733.

119. William Turnbull, *Reports on the Construction of the Piers of the Aqueduct of the Alexandria Canal across the Potomac at Georgetown, District of Columbia* (Washington, D.C.: U.S. Government Printing Office, 1873), pp. 10–11.

120. *American Quarterly Review* 32 (December 1834): 363.

121. Totten to Secretary of War Calhoun, April 17, 1821, Totten Papers, 1:81, NA, RG 77.

122. Long's design "was the first truss where mathematical calculations entered into the construction of the bridge"; see Wood, *Stephen Harriman Long*, p. 167.

123. Martin Reuss, "Andrew A. Humphreys and the Development of Hydraulic Engineering: Politics and Technology in the Army Corps of Engineers, 1850–1950," *Technology and Culture* 26 (January 1985): 2–3, and Hunter Rouse, *Hydraulics in the United States, 1776–1976* (Iowa City: University of Iowa Press, 1976), pp. 37–41.

124. Morison, *From Know-How to Nowhere*, p. 90.

4. Objects of National Pride

1. Stevenson, *Civil Engineering of North America*, pp. 19, 20.

2. George Roger Taylor, *The Transportation Revolution, 1815–1860* (New York: Rinehart & Co., 1951), p. 67.

3. Edward L. Pross, "A History of Rivers and Harbors Appropriation Bills, 1866–1933" (Ph.D. diss., Ohio State University, 1939), p. 41.

4. Carter Goodrich, ed., *The Government and the Economy, 1783–1861* (Indianapolis: Bobbs-Merrill Co., 1967), p. 44; see also Lawrence M. Friedman, *A History of American Law* (New York: Simon and Schuster, 1973), and Jamil Zainaldin, *Law in Antebellum Society: Legal Change and Economic Expansion* (New York: Alfred A. Knopf, 1983), pp. 19–21.

5. The spending estimate includes more than $1 million invested in canal company stock; about $2 million for piers, harbor repair, and coast surveying before 1824; $13 million for river and harbor work, 1824–1860; and $27 million for lighthouses, beacons, and buoys; see U.S. Treasury Department, "Statement of Appropriations and Expenditures," S. Exec. Doc. 196, 47th Cong., 1st sess. (1882), pp. 286, 521–522; see also Raymond H. Pulley, "Andrew Jackson and Federal Support of Internal Improvements: A Reappraisal," *Essays in History* 9 (1963–64): 48–59.

6. Alexander Hamilton, "Piers in Delaware River," H. Doc. 7, 5th Cong., 2d sess. (1798), pp. 1–2; see also Oliver Wolcott, "Piers in Delaware River," 4th Cong., 1st sess. (1796), p. 1, and Oliver Wolcott, "Tonnage Money Received Under the Act of March 17, 1800, and Applied to the Improvement of the Harbor of Baltimore and

Savannah River, February 28, 1822," *American State Papers, Commerce and Navigation* 2:626–627.

7. U.S. Constitution, Article 1, Section 8.

8. Archibald Cox, *The Court and The Constitution* (Boston: Houghton Mifflin Co., 1987), pp. 84–92; see also Carter Goodrich, "American Development Policy: The Case of Internal Improvements," *Journal of Economic History* 16 (December 1956): 449–460.

9. Joseph Hobson Harrison, Jr., "The Internal Improvement Issue in the Politics of the Union, 1783–1825," 2 vols. (Ph.D. diss., University of Virginia, 1954), 1:127; for the Jeffersonian position, see also John Lauritz Larson, "A Bridge, A Dam, A River: Liberty and Innovation in the Early Republic," *Journal of the Early Republic* 7 (Winter 1987): 351–375.

10. Henry G. Wheeler, *History of Congress: Biographical and Political*, 2 vols. (New York: Harper and Brothers, 1848), 2:146; see also Office of the Chief of Engineers, U.S. Army, "Laws of the United States Relating to the Improvement of Rivers and Harbors, vol. 1, 1790 to 1896," 62d Cong., 3d sess. (1940), pp. 18–19 [hereafter cited as "Laws Relating to Rivers and Harbors"].

11. Albert Gallatin, *Report of the Secretary of the Treasury, on the Subject of Public Roads and Canals* (Washington, D.C.: C. Weightman, 1808), pp. 6, 71.

12. Raymond Walters, Jr., *Albert Gallatin: Jeffersonian Financier and Diplomat* (Pittsburgh: University of Pittsburgh Press, 1969), pp. 181–184; see also Joseph H. Harrison, Jr., "*Sic et Non:* Thomas Jefferson and Internal Improvement," *Journal of the Early Republic* 7 (Winter 1987): 335–349, and Carter Goodrich, "The Gallatin Plan after One Hundred and Fifty Years," *American Philosophical Society Proceedings* 102 (1958): 436–441.

13. On the engineer's disdain for speculation, see Edwin T. Layton, Jr., *The Revolt of the Engineers: Social Responsibility and the American Engineering Profession* (Baltimore: Johns Hopkins University Press, 1966), p. 53.

14. *Memoirs of General Joseph Swift*, p. 49.

15. Cullum, *Campaigns of the War of 1812*, pp. 189–193; see also Edward M. Coffman, *The Old Army: A Portrait of the American Army in Peacetime, 1784–1898* (New York: Oxford University Press, 1986), p. 34.

16. Regarding Lee's support of internal improvements, biographer Douglas Southall Freeman says, "A Whig politician would hardly have been more sincere"; see Freeman, *R. E. Lee: A Biography*, 4 vols. (New York: Charles Scribner's Sons, 1949), 1:177; see also John W. Larson, "Ante-Bellum Channel and Harbor Improvements on the Great Lakes" (paper prepared for roundtable on the Corps of Engineer's North Central Division, n.d., Corps Office of History, Fort Belvoir, Va.), p. 14.

17. George Dangerfield, *The Era of Good Feelings* (Chicago: Ivan R. Dee, Inc., 1989), p. 95.

18. Henry Clay, *The Works of Henry Clay, Comprising his Life, Correspondence, and Letters*, ed. Calvin Colton, 7 vols. (New York: Henry Clay Publishing Co., 1897), 5:396; for the French influence on Clay's concept of the American System, see Merrill D. Peterson, *The Great Triumvirate: Webster, Clay, and Calhoun* (New York: Oxford University Press, 1987), p. 76.

19. Thomas Wilson, "Roads and Canals," H. Doc. 69, 14th Cong., 2d sess. (1817), pp. 6–7.

20. Henry St. George Tucker, "Roads and Canals," 15th Cong., 1st sess. (1818), p. 7.

21. U.S. Congress, *Annals of Congress,* ed. Joseph Gales and William Winston Seaton, 42 vols. (Washington, D.C: Gales and Seaton, 1834–56); see 14th Cong., 2d sess. (1817), pp. 296–297, 851–958.

22. John C. Calhoun, "Roads and Canals," H. Doc. 80, 16th Cong., 2d sess. (1819), p. 1; for Calhoun's thoughts on military roads during the Bonus Bill debate, see for example, *Annals of Congress,* 14th Cong., 1st sess. (1817), pp. 840–846.

23. Stephen H. Long, "Canal to Connect Illinois River and Lake Michigan," H. Doc. 81, 16th Cong., 1st sess. (1819), p. 4; see also Thomas Wilson, "Chesapeake and Delaware and Dismal Swamp Canals," H. Doc. 71, 14th Cong., 2d. sess. (1817), pp. 1–4.

24. *Annals of Congress,* 14th Cong., 2d sess. (1817), p. 1062.

25. John Lauritz Larson, "'Bind the Republic Together': The National Union and the Struggle for a System of Internal Improvements," *Journal of American History* 74 (September 1987): 392–385; for Madison's fears, see Gaillard Hunt, ed., *The Writings of James Madison,* 9 vols. (New York: G. P. Putnam's Sons, 1900–1910), 8:386–88.

26. "Laws Relating to Rivers and Harbors," pp. 21–22; see also James Monroe, "Message . . . on the Subject of Internal Improvements," H. Doc. 127, 17th Cong., 1st sess. (1822), pp. 7–60, and Simon Bernard and Joseph G. Totten, "Report of the Engineers . . . on the Subject of a Breakwater in the Delaware, July 14, 1823," in William Jones, *Remarks on the Proposed Breakwater at Cape Henlopen* (Philadelphia: Chamber of Commerce of Philadelphia, 1826), pp. 13–26.

27. Alvin F. Harlow, *Old Towpaths: The Story of the American Canal Era* (New York: D. Appleton and Co., 1926), p. 71.

28. U.S. Congress, Annals of Congress, 18th Cong., 1st sess. (1824), p. 3217.

29. James F. Hopkins, ed., *The Papers of Henry Clay,* vol. 3, *Presidential Candidate, 1821–1824* (Lexington: University of Kentucky Press, 1963), p. 583; see also U.S. Department of State, *Connected View of the Whole Internal Navigation of the United States* (Philadelphia: H. C. Carey & I. Lea, 1826), and, for a comprehensive review, Ronald E. Shaw, *Canals for a Nation: The Canal Era in the United States, 1790–1860* (Lexington: University Press of Kentucky, 1990).

30. "*Gibbons v. Ogden,*" in B. R. Curtis, comp., *Reports of Decisions in the Supreme Court of the United States* (Boston: Little, Brown, and Co., 1870), 6:2.

31. Maurice G. Baxter, *The Steamboat Monopoly: Gibbons v. Ogden, 1824* (New York: Alfred A. Knopf, 1972), pp. 40–41; see also W. Howard Mann, "The Marshall Court: Nationalization of Private Rights and Personal Liberty from the Authority of the Commerce Clause," *Indiana Law Journal* 38 (Winter 1963), pp. 117–238.

32. Curtis, *Decisions in the Supreme Court* 6:5, 8.

33. Donald J. Pisani, "Promotion and Regulation: Constitutionalism and the American Economy," *Journal of American History* 74 (December 1987): 742–744; see also Cox, *Court and Constitution,* pp. 86–88.

34. Wheeler, *History of Congress,* pp. 179, 181.

35. P. P. Barbour, "Roads and Canals," *Niles Register,* February 21, 1824, pp. 394, 396.

36. *Annals of Congress,* 18th Cong., 1st sess. (1824), pp. 534–551, 998–999; see also Hill, *Roads, Rails, and Waterways,* pp. 46–48, and, for the western position, R. B. Way, "The Mississippi Valley and Internal Improvements, 1825–1840," *Proceedings of the Mississippi Valley Historical Association* 4 (1910–1911): 53–180.

37. Harrison, "The Internal Improvement Issue," appendix B; see also *Annals of Congress,* 18th Cong., 1st sess. (1824), pp. 568–571, 1463, 1467–1469.

38. From the text of the act; see "Laws Relating to Rivers and Harbors," p. 26.

39. "Internal Improvements," *Niles Register,* April 9, 1825, pp. 89–95; see also "Corps of Engineers," *Niles Register,* October 32, 1825, pp. 121–122.

40. Carter Goodrich, "National Planning of Internal Improvements," *Political Science Quarterly* 63 (January 1941): 41.

41. Richardson, *Messages and Papers of the Presidents* 2:872, 882.

42. Dupree, *Science in the Federal Government,* pp. 39–43; see also Richard Hofstadter, *Anti-Intellectualism in American Life* (New York: Vintage Books, 1962), pp. 157–158.

43. Alexander Macomb, "Report of the operations of the Engineer Department, November 1, 1825," *ASP, MA* 3:137–138; see also "Internal Improvements," *North American Review,* n. s., 29 (January 1927): 1–23; for a list of projects, 1824 to 1828, see Charles Gratiot, "Statement showing the several works of Internal Improvement," in "Report from the Secretary of War," H. Doc. 7, 25th Cong., 2d sess. (1828), appendix 2; for rejected projects, see Macomb, "Surveys of Roads and Canals and of Their Relative Importance," in "Letter from the Secretary of War," H. Doc. 106, 19th Cong., 2d sess. (1927), pp. 5–6.

44. U.S. Board of Engineers, "Survey of Muscle Shoals," H. Doc. 284, 20th Cong., 1st sess. (1828), p. 18.

45. U.S. Board of Engineers, "Report of the U.S. Army Engineers on the Morris Canal," quoted in Goodrich, *Government and the Economy,* p. xvii.

46. Stephen H. Long, "Harbor of Delaware City, and the Navigation of Back Creek," H. Doc. 199, 25th Cong., 3d sess. (1830), p. 1.

47. U.S. Congress, *Register of Debates in Congress,* ed. Joseph Gales and William Winston Seaton, 14 vols. (Washington, D.C.: Gales and Seaton, 1824–1837); see 21st Cong., 1st sess. (1830), p. 1148; see also Ralph D. Gray, *The National Waterway: A History of the Chesapeake and Delaware Canal, 1769–1985,* 2d ed. (Urbana: University of Illinois Press, 1989), pp. 47–48.

48. George Washington Ward, *The Early Development of the Chesapeake and Ohio Canal Project* (Baltimore: Johns Hopkins University Press, 1899), p. 75; see also Abner Lacock, *Great National Object: Proposed Connection of the Eastern & Western Waters, by a Communication through the Potomac Country* (Washington, D.C.: n.p., 1822). The most complete history is Walter S. Sanderlin, *The Great National Project: A History of the Chesapeake and Ohio Canal* (Baltimore: Johns Hopkins University Press, 1946).

49. Jefferson to George Washington, March 15, 1784; quoted in Shaw, *Canals for a Nation,* p. 8.

50. Franklin, "The Tidewater End of the Chesapeake and Ohio Canal," p. 291; for Mercer and the procanal lobby, see *Proceedings at a General Convention . . . Held in the City of Washington . . . on the Subject of the Chesapeake and Ohio Canal* (Washington, D.C: Washington Republican, 1823); for the debate in the House, see *Niles Register,* February 21, 1824, pp. 393–397.

51. Wheeler, *History of Congress,* p. 183.

52. Totten to John Abert, May 15, 1824, Totten Papers, 1:56, NA, RG 77.

53. John J. Abert, Survey of the Potomac Canal, 1824 and 1825, 2 parts., 1:4, 2:22, no. 63, box 9, Bulky Package File, entry 249-A, NA, RG 77; see also Macomb to Secretary of War Calhoun, June, 13, 1824; Macomb to M. Rogers, Esq. [Pennsylvania] Secretary of State, Letters Sent Relating to Internal Improvements, 1824–1830, and Macomb to John Abert, June 14, 1824, entry 249, NA, RG 77.

54. For Bernard's strong endorsement, see "A Report of the Examination Which Has Been Made by the Board of Engineers," S. Doc. 32, 18th Cong., 2d Sess. (1825), p. 11; for the summit level, see William McNeill, "Descriptive Memoir . . . for a Canal to Connect the Youghangany [*sic*] & Potomac Rivers, 1825, no. 70, box 9, Bulky Package File, entry 292-A, NA, RG 77

55. Simon Bernard et al., "Report from the Board of Engineers . . . Concerning the Proposed Chesapeake and Ohio Canal," H. Exec. Doc. 10, 19th Cong., 2d sess. (1826), p. 66; for engineering debate over fluid dynamics and the importance of a sloped canal bed, see John MacNeill, *Canal Navigation; on the Resistance of Water to the Passage of Boats upon Canals, and Other Bodies of Water, Being the Result of Experiments* (London: Roake and Varty, 1833), p. 1.

56. Bernard, "Report Concerning the Chesapeake and Ohio Canal," p. 23. When work began in 1828, the company paid nonskilled labor only about ten dollars a month (with a promise of meals and liquor).

57. Charles Mercer, "Chesapeake and Ohio Canal," H. Rept. 141, 20th Cong., 1st sess. (1828), p. 8; see also James Barbour, "Letter from the Secretary of War Transmitting Estimates of the Cost of Making a Canal," H. Doc. 192, 20th Cong., 1st sess. (1828), pp. 98–99.

58. Joseph H. Harrison, Jr., "Simon Bernard, the American System, and the Ghost of the French Alliance," in John B. Boles, ed., *America: The Middle Period; Essays in Honor of Bernard Mayo* (Charlottesville: University Press of Virginia, 1973), pp. 145–167; see also William H. Carter, "Bvt. Maj. Gen. Simon Bernard," *Professional Memoirs* 5 (May–June 1913): 306–314, and "General Simon Bernard: Aide to Napoleon, Designer of Fort Monroe," *Tales of Old Fort Monroe* 4 (Fort Monroe, Va.: Fort Monroe Casemate Museum, n.d.); for Bernard's Atlantic passage, see Hunter and Dooley, *Claudius Crozet*, p. 15.

59. Harrison, "Simon Bernard and the French Alliance," pp. 149, 151.

60. Hill, *Roads, Rails, and Waterways,* p. 7.

61. Joseph Swift thought Bernard "a cold-hearted man; not in any sense a genius"; see *Memoirs of General Joseph Swift,* p. 180.

62. Asbel Welsh, president of the American Society of Civil Engineers, quoted in Harlow, *Old Towpaths,* p. 299.

63. Charles B. Stuart, *Lives and Works of Civil and Military Engineers* (New York: D. Van Nostrand, 1871), pp. 109, 118.

64. Roberts, known primarily for his staircase locks at Lockport, New York, was also the mentor of practical engineers like John B. Jervis; see Ronald E. Shaw, *Erie Water West: A History of the Erie Canal, 1792–1854* (Lexington: University of Kentucky Press, 1966), p. 89; see also John A. Krout, "New York's Early Engineers," *New York State Historical Society Quarterly Journal* 26 (July 1945): 269–277.

65. Mercer, "Chesapeake and Ohio Canal," H. Rept. 141, 20th Cong., 1st sess. (1828), p. 3.

66. "Memorial of the Chesapeake and Ohio Canal Company," H. Doc. 12, 20th Cong., 2d sess. (1828), p. 18; see also *National Intelligencer*, May 27, 1828, p. 1, and Sanderlin, *Great National Project*, pp. 56–68.

67. *Sketch of the Geographical Rout* [sic] *of a Great Railway* (New York: G. C. & H. Carvill, 1830), p. 16.

68. "Memorial of the President and Director of the Chesapeake and Ohio Canal [relative to the Baltimore and Ohio Railroad]," S. Doc. 99, 20th Cong., 2d sess. (1829); see also Hugh Sisson Hanna, *A Financial History of Maryland (1789–1848)* (Baltimore: Johns Hopkins University Press, 1907), pp. 79–96.

69. John J. Abert and James Kearney, Report on . . . Portions of the Chesapeake and Ohio Canal . . . , June 18, 1831, no. 7a, box 10, p. 14, Bulky Package File, entry 292-A, NA, RG 77; see also William Gibbs McNeill, *Report from Gen. Wm. G. McNeill, Late President of the Chesapeake and Ohio Canal Company, to the Governor of Maryland, December 1843* (Annapolis, 1844), William Henry Swift, *Report on the Present State of the Chesapeake and Ohio Canal* (Boston: Dutton and Wentworth, 1846), and James Barbour, "Letter of Secretary of War . . . on Survey of Route for Proposed Canal to Connect the Chesapeake and Ohio with Baltimore," H. Exec. Doc. 58, 20th Cong., 1st sess. (1828).

70. Chesapeake and Ohio Canal Company, *Report to the Stockholders on the Completion of the Chesapeake and Ohio Canal to Cumberland . . . February, 1851* (Frederick, Md.: D. Schley and T. Haller, 1851), pp. 39, 82–83; for construction estimates, see also Archer Butler Hulbert, *The Great American Canals*, 2 vols. (New York: AMS Press, 1971), vol. 1, *The Chesapeake and Ohio Canal and the Pennsylvania Canal*, pp. 160–163.

71. Bernard, "Report Concerning the Chesapeake and Ohio Canal," p. 23.

72. Mercer, "Chesapeake and Ohio Canal," H. Rept. 141, 20th Cong., 1st sess. (1828), pp. 4, 6.

73. Ward, *Early History of the Chesapeake and Ohio Canal Project*, p. 75.

74. In 1828 Congress had insisted on a 60-foot width at least from Little Falls to Georgetown. The canal company, at Bernard's suggestion, built to a width of 60-feet and a depth of at least 6 as far as Harpers Ferry, but financial constraints above that point forced management to abandon these dimensions. One bottleneck was the 3,118-foot Paw Paw tunnel in the mountains above Williamsport. Blasted through slate, it narrowed the channel to 17 feet. Locks and aqueducts remained 15 feet throughout; see Shaw, *Canals for a Nation*, pp. 104–105; see also "Laws Relating to Rivers and Harbors," p. 48.

75. Topographer James Kearney, however, was Irish-born; for Bernard's resignation, see Francoise Planchot, "Le General Simon Bernard, ingenieur militaire aux Etats-Unis," *Revue Française D'Etudes Américaines* 13 (February 1982): 95.

76. John J. Abert, "Topographical Engineers," H. Rept. 95, 24th Cong., 1st sess. (1835), p. 14.

77. K. Jack Bauer, *A Maritime History of the United States: The Role of America's Seas and Waterways* (Columbia: University of South Carolina Press, 1988), pp. xi, 144; for rivers and the ecomomic growth of the Ohio Valley, see Erik F. Haites, James Mak, and Gary M. Walton, *Western River Transportation* (Baltimore: Johns Hopkins University Press, 1975).

78. River hazards are classified in Samuel S. Forman, *Narrative of a Journey Down the Ohio and Mississippi in 1789–90* (Cincinnati: H. W. Derby and Co., 1848); for the hazards of the Ohio, see, for example, James Flint, *Letters from America* (Edinburgh: W. & C. Tait, 1822), p. 260, passim.; for snag accidents, see Louis C. Hunter, *Steamboats on the Western Rivers: An Economic and Technological History* (Cambridge: Harvard University Press, 1949), pp. 272–274.

79. Simon Bernard and Joseph G. Totten, "A Report of the Board of Engineers on the Ohio and Mississippi River . . . Made in the Year 1821," H. Doc. 35, 17th Cong., 2d sess. (1823), p. 21.

80. *Annals of Congress,* 18th Cong., 1st sess. (1842), pp. 2578–2585; see also Curtis Nettels, "The Mississippi Valley and the Constitution, 1815–1829," *Mississippi Valley Historical Review* 11 (December 1924): 332–338, and Charles Henry Ambler, *A History of Transportation in the Ohio Valley* (Glendale, Calif.: Arthur H. Clark Co., 1932), pp. 395–396.

81. Macomb to Stephen Long, June 8, 1824, and February 11, 1825, Letters Sent Relating to Internal Improvements, 1824–1830, entry 249, NA, RG 77; see also *DAB,* s.v. "Macomb, Alexander."

82. Alexander Macomb, "A Report of the Chief Engineer, Relative to the Application of the Appropriation for Removing Obstructions to the Navigation of the Ohio and Mississippi Rivers," S. Doc. 14, 19th Cong., 1st sess. (1826), p. 7.

83. Macomb to Samuel Babcock, November 16, 1824, Letters Sent Relating to Internal Improvements, 1824–1830, entry 249, NA, RG 77; for the snag boat contest, see Leland R. Johnson, "19th Century Engineering, Part 1: The Contest of 1824," *Military Engineer* 65 (May–June 1973): 66–71.

84. U.S. Congress, "Survey of the Works at Fort Delaware, and Proceedings of a Court-Martial of Major Babcock," *ASP, MA* 2:792; see also Johnson, *Falls City Engineers,* pp. 50–55.

85. Babcock to Alexander Macomb, October 3, 1825, in Macomb, "A Report . . . Relative to . . . the Ohio and Mississippi Rivers," S. Doc. 14, 19th Cong., 1st sess. (1826), pp. 11–12.

86. William M. Poyntz to Henry Clay, November 1, 1826, in ibid., p. 14.

87. "Laws Relating to Rivers and Harbors," p. 27.

88. Macomb, "A Report . . . Relative to . . . the Ohio and Mississippi Rivers," p. 14.

89. Macomb was an expert on court-martial, having authored an 1809 treatise on the subject; see Cullum, *Biographical Register,* 1:57–58; see also "Petition of John Bruce for remuneration of account of the losses on a contract for removing obstructions in the Ohio and Mississippi Rivers," S. Doc. 421, 26th Cong., 1st sess. (1840), pp. 1–2.

90. Gary B. Mills, *Of Men and Rivers: The Story of the Vicksburg District* (Vicksburg, Miss.: U.S. Army Engineer District, 1978), p. 19.

91. Steamboat critic Joseph Hough, quoted in Michael Allen, "The Ohio River: Artery of Movement," in Robert L. Ried, ed., *Always a River: The Ohio River and the American Experience* (Bloomington: Indiana University Press, 1991), p. 117.

92. Long to Alexander Macomb, March 28, 1826, Letters Received, 1826–1866, Office of the Chief of Engineers; entry 18, NA, RG 77.

93. For heroic treatment, see J. Fair Hardin, "The First Great Western River Captain: A Sketch of the Career of Captain Henry Miller Shreve," *Louisiana Historical*

Quarterly 10 (1927): 25–67; see also Florence L. Dorsey, *Master of the Mississippi: Henry Shreve and the Conquest of the Mississippi* (Boston: Houghton Mifflin, 1941); for critical reappraisal, see Louis C. Hunter, "The Invention of the Western Steamboat," *Journal of Economic History* 3 (1943): 201–220.

94. Corps historian Charles Hendricks has discovered that Shreve named his boat "Helepolis" (Greek for "siege machine"), although very soon the "Heliopolis" spelling was common in Corps reports.

95. Shreve explained his methods in letters to Secretary of War James Barbour, February 20 and April 26, 1827, LRSW, Entry 18, NA, RG 77; for the *Heliopolis* trial at Plum Point, see Johnson, *Headwaters District,* pp. 72–73.

96. Alexander Bowman to Charles Gratiot, January 31, 1835, in "Report from the Engineer Department," S. Doc. 1, 24th Cong., 1st sess. (1835), p. 175.

97. "Memorial of Sundry Citizens of Louisville, Kentucky," H. Rpt. 337, 21st Cong., 1st sess. (1830); see also C. A. Wickliffe, "Navigation Ohio and Mississippi Rivers," H. Rept. 379, 21st Cong., 1st sess. (1830).

98. Hunter, *Steamboats on the Western Rivers,* pp. 199, 644; see also Clark Kittrell, "Navigation Improvement at Chain of Rocks," *Military Engineer* 40 (December 1948): 556–558; for President Jackson's ambivalent stand on river construction, see Edward Campbell Mason, *The Veto Power: Its Origins, Development and Function in the Government of the United States, 1789–1889* (Boston: Ginn and Co., 1830), pp. 97–99; for tree cutting, see Henry Shreve, "Navigation of the Mississippi River," H. Doc. 11, 20th Cong., 1st sess. (1827), pp. 4–5.

99. Charles Gratiot, "Report from the Engineer Department" S. Doc. 1, 24th Cong. (1835), pp. 168–175; see also Norman W. Caldwell, "The Red River Raft," *Chronicles of Oklahoma* 19 (1941): 253–268, and Harry Sinclair Drago, *Red River Valley: The Mainstream of Frontier History from the Louisiana Bayous to the Texas Panhandle* (New York: Clarkson N. Potter Co., 1962), pp. 97–98.

100. Dorsey, *Master of the Mississippi,* pp. 175–178.

101. Henry Shreve, "Red River," in Charles Gratiot, "Report from the Engineer Department," S. Doc. 1, 25th Cong., 3d sess. (1838), pp. 307–315; see also Mills, *Of Men and Rivers,* pp. 20–22.

102. Stephen Long, "A Report on the Improvement of the Red River," S. Doc. 64, 27th Cong., 1st sess. (1841); see also Ella Huchison Ellwanger, "Famous Steamboats and Their Captains on Western and Southern Waters," *Register of the Kentucky State Historical Society* 18 (January and May 1920): 25–26, and Johnson, *Falls City Engineers,* p. 86.

103. Snyder and Guss, *The District,* p. 64.

104. "The Invention of the Dredging Machine," *Mechanics Magazine* (London) 51 (1849?): 126.

105. David F. Bastian, "The Development of Dredging through the 1850s," in *National Waterway Roundtable Papers: Proceedings on the History of Evolution of U.S. Waterways and Ports* (Norfolk, Va.: U.S. Army Engineer Institute for Water Resources, 1980), pp. 1–21; see also Arthur E. Francke, Jr., "Early Corps River Dredging in Florida, 1829–40" (Manuscript prepared for the U.S. Corps of Engineers History Office [typewritten 1990]), pp. 15–19.

106. "Laws Relating to Rivers and Harbors," pp. 46–48; see also Isaac Roberdeau, Report on Mud Machine, April 8, 1822, Roberdeau Papers, entry 306,

NA, RG 77; for ladder dredging, see David B. Macomb, "Survey and Estimate for the Improvement of the Harbor and River of St. Mark's in Florida," H. Doc. 76, 20th Cong., 2d sess. (1829), p. 4; for harrow scraping, see J. R. Putnam, *Plan for Removing Bars at the Mouth of the Mississippi River and Other Harbors* (New Orleans: Bulletin Office, 1841).

107. Walter M. Lowrey, "Navigational Problems at the Mouth of the Mississippi River, 1698–1880" (Ph.D. diss., Vanderbilt University, 1956), p. 129.

108. Harry E. Barber, "The History of the Florida Cross-State Canal" (Ph.D. diss., University of Georgia, 1969), pp. 36–43.

109. *Register of Debates in Congress*, 24th Cong., 1st sess. (1836), pp. 1934–1395.

110. "Laws Relating to Rivers and Harbors," pp. 70–72, 76–79; see also Lowrey, "Navigational Problems at the Mouth of the Mississippi River," pp. 123–124.

111. Although capped at twenty-two engineers, army registers list twenty-five Corps engineers on active duty in 1837; see Andrew A. Humphreys, "Historical Papers Relating to the Corps of Engineers and to Engineer Troops in the United States Army," *Occasional Papers*, U.S. Army Engineer School, 16 (1904): 24, 47.

112. John J. Abert, "Topographical Engineers," H. Rept. 95, 24th Cong., 1st sess. (1834), p. 2.

113. John J. Abert, "A report in relation to the Coast Survey," H. Doc. 13, 24th Cong., 2d sess. (1836), p. 16; see also *DAB*, s.v. "Abert, John James"; for Abert's lobbying on behalf of the bureau, see for example, Abert to Peter Porter, February 12, 1829, Letters Sent, 1829–1867, Records of the Topographical Bureau, Entry 310, NA, RG 77.

114. Livingston, "Col. Stephen H. Long," 4:490.

115. Stephen H. Long, "Improvement of the Holston and Tennessee Rivers [September 1832]," H. Exec. Doc. 167, 43d Cong., 2d sess. (1875), pp. 19, 20; see also John J. Abert, "Survey of the Kennebec River," H. Doc. 108, 19th Cong., 2d sess. (1827), p. 6; for the topographical bureau and river dredging, see Ronald Tweet, *A History of the Rock Island District, Corps of Engineers, 1866–1975* (Rock Island, Ill.: U.S. Army Engineer District, 1975), pp. 24–25.

116. Abert to Peter B. Porter, February 12, 1829, Letters Sent, 1829–1867, Records of the Topographical Bureau, entry 310, NA, RG 77.

117. Recruits of 1838 with second careers in Congress were George W. Hughes, Robert E. McLane, and Joseph E. Johnson. Hughes also became a railroad president as did William Swift. Topographers who became place-names include Long, James D. Graham, John C. Frémont, William H. Emory, Howard Stansbury, John W. Gunnison, William H. Warner, Lorenzo Stitgreaves, and Andrew A. Humphreys. Abert left his name to a steamboat, a western species of squirrel, and a lake in South Dakota (renamed Lake Albert); see Cullum, *Biographical Register* 1:passim; for place-names, see also Frank N. Schubert, *Vanguard of Expansion: Army Engineers in the Trans-Mississippi West, 1819–1879* (Washington, D.C.: U.S. Army Corps of Engineers Historical Division, 1980), p. 149, and Bray, *Joseph Nicollet and His Map*, p. 202.

118. *Congressional Globe* 6 (January 24, 1838): 133–134.

119. Circular, July 11, 1838, Letters Sent, 1829–1867, Records of the Topographical Bureau, entry 310, NA, RG 77; see also Raphiel P. Thian, comp., *Legislative History of the General Staff of the Army of the United States . . . from 1775 to 1901* (Washington,

D.C.: U.S. Government Printing Office, 1901), pp. 502–503, and Ryan, "War Department Topographical Bureau," p. 124.

120. Joseph Hobson Harrison, Jr., "Martin Van Buren and His Southern Supporters," *Journal of Southern History* 22 (November 1, 1956): 456.

121. Goodrich, *Government Promotion of American Canals and Railroads,* pp. 43–48; see also Emory R. Johnson, "River and Harbor Bills," *Annals of the American Academy of Political and Social Science* 2 (1892): 788–90.

122. Goodrich, *Government Promotion of American Canals and Railroads,* p. 17.

123. Francis E. Rourke, *Bureaucracy, Politics, and Public Policy* (Boston: Little, Brown, and Co., 1984), p. 15.

124. Hill, *Roads, Rails, and Waterways,* p. 192.

125. Hofstadter, *Anti-Intellectualism in American Life,* p. 155.

126. Leonard D. White, *The Jacksonians: A Study in Administrative History, 1825–1861* (New York: The Macmillan Co., 1954), and White, *The Jeffersonians: A Study in Administrative History, 1801–1829* (New York: The Macmillan Co., 1951); see also Matthew A. Crenson, *The Federal Machine: Beginnings of Bureaucracy in Jacksonian America* (Baltimore: Johns Hopkins University Press, 1975), pp. 5–8.

127. Ellis and Moore, *School for Soldiers,* p. 193.

128. Eventually Strong won an appointment by making his case in person to President Franklin Pierce; see George C. Strong, *Cadet Life at West Point* (Boston: T.O.H.P. Burnham, 1862), pp. 33, 35–37, 74.

129. Totten even coauthored the 1852 aqueduct funding bill; see Albert E. Cowdrey, *City for the Nation: The Army Engineers and the Building of Washington, D.C., 1790–1967* (Washington, D.C.: U.S. Army Engineers Historical Division, 1979), pp. 16–17; for a good hint at Abert's politics, see Goetzmann, *Army Exploration in the American West,* p. 10. Joseph Swift fit the classic profile of the antebellum Whig—an educated professional from a Federalist family, an ardent member of the Episcopal church, a moderate antislavery man who supported the Compromise of 1850; see Swift, *Memoirs;* Thayer's biographer calls him "a staunch New England conservative" who supported Clay and Fillmore before joining Frémont's Republicans in 1856; see Kershner, "Sylvanus Thayer," p. 293.

130. John J. Abert, "Report from the Topographical Bureau," H. Doc. 2, 27th Cong., 3d sess. (1842), p. 277; see also Abert, "Report from the Chief of Topographical Engineers," S. Doc. 1, 20th Cong., 1st sess. (1845), pp. 297–299, and Ryan, "War Department Topographical Engineers," pp. 136–142.

5. "A Privileged Order of the Very Worst Class"

1. *The Military Academy at West Point, Unmasked; or, Corruption and Military Despotism Exposed, by Americanus* [psued. Alden Partridge] (Washington, D.C.: n.p., 1930), pp. 1, 3, 18–20; for Partridge and the attack on West Point, see also Arthur A. Ekirch, Jr., *The Civilian and the Military* (New York: Oxford University Press, 1956), pp. 69–71, and R. L. Watson, "Congressional Attitudes toward Military Preparedness, 1829–1835," *Mississippi Valley Historical Review* 34 (March 1948): 611–636.

2. Hofstadter, *Anti-Intellectualism in American Life*, pp. 162–163.

3. Taylor to Gen. Thomas S. Jesup, September 18, 1820; quoted in Holman Hamilton, *Zachary Taylor* (Indianapolis: Bobbs-Merrill, 1941), p. 67.

4. "Resolutions of the State of Tennessee Demanding the Abolition of West Point, November, 1833," *ASP, MA* 7:89. Caleb Atwater refers to the aristocratic stereotype in a letter to Colonel Abert, July 15, 1850, Records of the Topographical Bureau, 1818–1867, Letters Received, 1832–1865, entry 315, NA, RG 77; see also John J. Lenny, *Corps of Engineers and Their West Point System* (New York: Greenberg, 1949), pp. 196–197.

5. Denton, "The Formative Years of the United States Military Academy," p. 249.

6. "Applications and Appointments as Cadets at the Military Academy at West Point from Its Establishment to 1829," *ASP, MA* 4:307; see also *Niles Register* (January 23, 1830), p. 374; Hamilton Fish, "Military Academy," H. Rept. 476, 28th Cong., 1st sess. (1844), p. 15, passim; and Joseph Totten, "Report of the Chief Engineer," S. Exec. Doc. 1, 29th Cong., 1st sess. (1845), p. 271.

7. Freeman, *R. E. Lee*, p. 43.

8. Joseph Totten, "Report of the Chief Engineer," S. Exec. Doc. 1, 29th Cong., 1st sess. (1845), p. 271; for "influence at elections" quote, see the West Point Board of Visitor's minority report in Joseph G. Totten, "Report from the Chief Engineer," S. Doc. 2, 26th Cong., 2d sess. (1840), p. 150.

9. Francis O. Smith, "History of the Origin of the Military Academy, 1837," *ASP, MA* 7:13; see also, *Niles Register*, January 23, 1830, p. 374, and T. Harry Williams, *P. G. T. Beauregard: Napoleon in Gray* (New York: Collier Books, 1955), pp. 16–201. James Morrison claims the academy did not cater to blue bloods; see *The Best School in the World*, p. 62.

10. Francis O. Smith, "Harbors and Rivers," H. Rept. 297, 24th Cong., 1st sess. (1836), pp. 1–3.

11. Richardson, *Messages and Papers of the Presidents* 4:1518–1519.

12. Erik McKinly Eriksson, "The Federal Civil Service under President Jackson," *Mississippi Valley Historical Review* 13 (March 1927): 519, 529.

13. Sen. John J. Crittenden's address on army expansion, July 4, 1838, in *Congressional Globe* 6:490; for the antipower critique of the capitol city, see James Sterling Young, *The Washington Community, 1800–1828* (New York: Harcourt, Brace & World, 1966), pp. 50, 57.

14. Richardson, *Messages and Papers of the Presidents* 3:1012.

15. Hill, *Roads, Rails, and Waterways*, p. 178.

16. Cullum, *Campaigns of the War of 1812*, pp. 350–352; see also Charles Hendricks, "The Dismissal of Chief Engineer Charles Gratiot," *Engineer Update* 12 (December 1988): 7.

17. William L. Jenks, "Fort Gratiot and Its Builder Gen. Charles Gratiot," *Michigan History Magazine* 4 (January 1920): 141–155; see also George T. Ness, "Missouri at West Point: Her Graduates through the Civil War Years," *Missouri Historical Review* 72 (April 1978): 162–169, and Simon Bernard, "Revised Report of the Board of Engineers on the Defense of the Seaboard, March 24, 1826," *ASP, MA*, 3:291.

18. Clary, *Fortress America*, p. 48.

19. Joel R. Poinsett, "General Charles Gratiot," H. Doc. 77, 25th Cong., 3d sess. (1839), p. 1; see also Curtis, *Decisions in the Supreme Court* 14:106–114.

20. In 1825 Gratiot's annual base salary (excluding bonuses) was $2,062. During eight years as disbursing officer at Hampton Roads, 1821 to 1829, he claimed an additional $5,758 ($719 a year) for each of the two fort assignments. General Bernard's annual salary was $3,090. According to historian Mark Paid, the typical civil engineer "of the first rank" could expect to make between $2,675 and $2,911, although by 1834 a chief engineer of a very big project might make up to $6,000 a year; see William Lee, "Brevet Rank and Pay," H. Doc. 41, 19th Cong., 2d sess. (1827), p. 13; see also Thomas M. Exley, *A Compendium of the Pay of the Army from 1785 to 1888* (Washington, D.C.: U.S. Governmment Printing Office, 1888), pp. 28–29, 59, and Mark Paid, "Earnings of American Civil Engineers, 1820–1859," *Journal of Economic History* 31 (June 1971): 407–419; for the argument that the Corps was underpaid, see Joseph Totten, et al., "Memorial . . . Against the Passage of the bill to regulate the pay and emoluments of the officers of the line and staff of the army," S. Doc. 376, 26th Cong., 1st sess. (1840), pp. 1–5. For scandals and plunder during the Jackson years, see Carl Russell Fish, *The Civil Service and the Patronage* (New York: Lohgmans, Green & Co., 1905), p. 114, passim.

21. The 1852 Senate Judiciary Report is reprinted in Cullum, *Biographical Register* 1:71.

22. Henry Clay, Jr. to Henry Clay, Sr., July 12, 1830, in *Papers of Henry Clay,* ed. Robert Seager II, vol. 8, *Candidate, Compromiser, Whig, 1829–1836* (Lexington: University Press of Kentucky, 1984), p. 235; for the fear of French-style centralization, see also civil engineer Coleman Sellar's comments quoted in Bruce Sinclair, "The Direction of Technology," in Edwin T. Layton, Jr., ed., *Technology and Social Change in America* (New York: Harper & Row, 1973), p. 77.

23. Weigley, *History of the United States Army,* pp. 177–178.

24. Diane Lindstrom, *Economic Development in the Philadelphia Region, 1810–1850* (New York: Columbia University Press, 1978), p. 37.

25. Simon Bernard et al., "Report . . . respecting a breakwater in the Delaware Bay, Philadelphia, July 14, 1823," Joint Report of the Secretary of Navy and War, item 161, Bulky Package File, E 292A, NA, RG 77; see also "Memorial of the Chamber of Commerce of the City of Philadelphia upon the Construction of a Breakwater in the Bay of Delaware," Sen. Doc. 5, 19th Cong., 1st sess. (1825), pp. 4–6, and Philadelphia Chamber of Commerce, "Cases of Shipwreck, Loss, and Disaster within the Bay of Delaware," H. Doc. 26, 19th Cong., 2d sess. (1826), pp. 3–10.

26. William Jones, *Remarks on the Proposed Breakwater at Cape Henlopen,* 2d ed. (Philadelphia: Chamber of Commerce of Philadelphia, 1826), pp. 22, 25–26; see also Joseph K. F. Mansfield, "Report from the Engineer Department, November 30, 1836," *ASP, MA* 6:879–888; for the French theory of waves, see J. F. Lane, "Delaware Breakwater," H. Doc. 145, 23d Cong., 2d sess. (1835), p. 7.

27. Thomas S. Jesup, "Delaware Breakwater—Names and Compensation of Superintendents," H. Doc. 422, 23d Cong., 1st sess. (1834), p. 2.

28. John G. Barnard, "Delaware Breakwater," H. Rept. 353, 34th Cong., 1st sess. (1856), p. 7.

29. Snyder and Guss, *The District,* pp. 14, 71–73; for breakwater appropriations, see C. W. Raymond, "Analytical and Topical Index to the Reports of the Chief of Engineers," H. Doc. 439, 57th Cong., 2d sess. (1903), pp. 311–315.

30. D. H. Mahan, *Elementary Course of Civil Engineering,* 1851 ed., p. 349.

31. Smith, "Harbors and Rivers," pp. 1–2, 31, 39.

32. Gratiot responded to Smith's allegations in Lewis Cass, "Works of Internal Improvement," H. Doc. 212, 24th Cong., 1st sess. (1836), pp. 1–56; see also Charles Gratiot, "Estimates to Close Accounts in Relation to Sundry Public Works," H. Doc. 767, 24th Cong., 1st sess. (1836), pp. 1–11.

33. Arthur L. Bates, *Remarks before the River and Harbor Committee in Behalf of Erie Harbor* (Washington, D.C.: n.p., 1902), p. 2.

34. Smith Thompson, "Obstructions to Navigation in the Harbor of Presque Isle, January 19, 1820," *American State Papers, Commerce and Navigation* 2:417; see also "Harbor of Erie," S. Doc. 95, 16th Cong., 2d sess. (1821), p. 1, and Geoffrey Perret, *A Country Made by War: From the Revolution to Vietnam—the Story of America's Rise to Power* (New York: Random House, 1989), pp. 117–118.

35. Drescher, *Engineers for the Public Good*, p. 71; see also U.S. House, "Contract for Stone at the Rip Raps at Old Point Comfort, May 7, 1822," *ASP, MA* 2:431–449.

36. Theodore Maurice to Alexander Macomb, October 22, 1826, Letters Received, Office of the Chief of Engineers, 1826–1866, entry 18, NA, RG 77.

37. Maurice to Alexander Macomb, November 12, 1826, and Theodore Maurice, Report of the Condition of the United States Works of Internal Improvement for Purposes of Navigation on Lake Erie, September 30, 1827, Letters Received, Office of the Chief of Engineers, 1826–1866, entry 18, NA, RG 77; see also Charles Gratiot, "Expenditures—Harbor of Presque Isle," H. Doc. 76, 23d Cong., 2nd sess. (1835), pp. 1–4.

38. Smith, "Harbors and Rivers," p. 29.

39. Thomas Brown, "Harbor of Presque Isle," H. Doc. 128, 24th Cong., 1st sess. (1836), pp. 2–6; for thoughts on the science of tides, see Brown to Charles Gratiot, May 30, 1838, Letters Sent by the Office of Harbor Improvements On Lake Erie, 1835–1837, entry 332, NA, RG 77.

40. "Memorial of the Town Council and Citizens of Erie, Pa.," S. Doc. 23, 26th Cong., 2d sess. (1841), p. 2.

41. Larson, "Ante-Bellum Channel and Harbor Improvements on the Great Lakes," p. 12.

42. John J. Abert, "Report of the Chief of the Bureau of Topographical Engineers, November 15, 1941," S. Doc. 1, 27th Cong., 2d sess, (1841), p. 196; see also Abert, "Report . . . relative to harbors between Erie and Buffalo," S. Doc. 358, 29th Cong., 1st sess. (1946), pp. 2–4, and Mentor L. Williams, "The Background of the Chicago Harbor and River Convention, 1847," *Mid-America* 30 (October 1948): 222.

43. Ryan, "War Department Topographical Bureau," p. 124.

44. House Committee on Commerce Report on Lighthouses, May 25, 1842; quoted in U.S. Light-House Establishment, *Compilation of Public Documents and Extracts from Reports and Papers Relating to Light-Houses*, p. 334.

45. *U.S. Statutes at Large* 5:661; for a breakdown of expenditures, see Johnson, "River and Harbor Bills," pp. 782–812.

46. Williams to John J. Abert, November 13, 1839, Letters Sent, U.S. Lake Survey, 1839–1845, Entry 348, NA, RG 77.

47. Williams to J. J. Barker, Esq., Chairman of the Committee in Charge of Harbor Improvements, December 4, 1842, Letters Sent, U.S. Lake Survey, 1839–1845, entry 348, NA, RG 77; for scientific accomplishments, see U.S. Lake Survey, *The U.S. Lake Survey Story* (Detroit, 1972).

48. Wheeler, *History of Congress*, p. 446.

49. *Congressional Globe*, 29th Cong., 1st sess. (1846), p. 1181.

50. Mentor L. Williams, "The Chicago River and Harbor Convention, 1847," *Mississippi Valley Historical Review* 35 (March 1949): 610.

51. Williams, "Chicago River and Harbor Convention," p. 232; for a southern view, see James D. B. De Bow, "Chicago and Memphis Conventions," *De Bow's Review* 4 (September 1847): 123.

52. Williams, "Chicago River and Harbor Convention," p. 617.

53. John J. Abert, "Report of the Chief of the Corps of Topographical Engineers," S. Doc. 1, 29th Cong., 1st sess. (1845), pp. 297–298.

54. Williams, "Chicago River and Harbor Convention, 1847," pp. 621–622; see also Mentor L. Williams, "'A Shout of Derision': A Sidelight on the Presidential Campaign of 1848," *Michigan History* 35 (1949): 66–77, and Victor L. Albjerg, "Internal Improvements without a Policy, 1780–1861," *Indiana Magazine of History* 28 (June 1932): 176.

55. Clary, *Fortress America*, pp. 44, 69.

56. Denton, "The Formative Years of the United States Military Academy," p. 244.

57. "Memorial of Alden Partridge, 1841," in Henry Barnard, *Military Schools and Courses of Instruction in the Science and Art of War* (New York: Greenwood Press, 1969), [originally published in 1872] p. 851.

58. Caleb Atwater to John J. Abert, July 15, 1850, Letters Received, 1832–1865, Records of the Topographical Bureau, entry 315, NA, RG 77; see also Denton, "The Formative Years of the United States Military Academy," p. 246, and *DAB*, s.v. "McNeill, William Gibbs"; for the reduction of faculty pay, see Hamilton Fish, "Military Academy," H. Rept. 476, 28th Cong., 1st sess. (1844), pp. 21–22.

59. Stephen W. Sears, *George B. McCellan: The Young Napoleon* (New York: Ticknor and Fields, 1988), p. 11.

60. Joseph G. Totten, "Report of the Chief Engineer," H. Doc. 1, 30th Cong., 1st sess. (1847), p. 628; see also Ambrose, *Duty, Honor, Country*, p. 140, Douglas D. Martin, "The Indefatigables: Army Engineers in the War with Mexico, 1845–1848," U.S. Army Corps of Engineers Office of History (typewritten, 1977), p. 125, and Guy C. Swan III et al., "Scott's Engineers," *Military Review* 63 (March 1983): 61–69; for sappers and miners see Gustavus W. Smith, "Company 'A' Corps of Engineers, in the Mexican War," *Occasional Papers*, U.S. Army Engineer School, 16 (1904).

61. Schubert, *Nation Builders*, p. 59.

62. Carroll Foster Reynolds, "The Development of Military Floating Bridges by the United States Army" (Ph.D. diss., University of Pittsburgh, 1950), pp. 19–20.

63. Adrian George Traas, "The U.S. Army Topographical Engineers in the Mexican War" (M.A. thesis, Texas A&M University, 1971), p. 44; see also Goetzmann, *Army Exploration in the West*, pp. 123–127, 151, and Martin, "The Indefatigables," p. 130.

64. William S. McFeeley, *Grant: A Biography* (New York: W. W. Norton, 1982), p. 28.

65. Walter Millis, *Arms and Men: A Study of American Military History* (New Brunswick, N.J.: Rutgers University Press, 1981), p. 107.

66. John J. Abert, "Report of the Chief, Topographical Engineers," S. Exec. Doc. 1, 30th Cong., 1st sess. (1848), p. 325.

67. *Congressional Globe,* 30th Cong., 1st sess. (1848), appendix, pp. 709, 711.

68. Edwin Hale Abbot, A.M., *A Review of the Report upon the Physics and Hydraulics of the Mississippi River* (Boston: Crosby and Nichols, 1862), p. 28. This reviewer was the brother of the Delta survey's coauthor, Henry Larcom Abbot.

69. John J. Abert to Secretary of War Charles M. Conrad, January 19, 1852, in S. Exec. Doc. 49, 32d Cong., 1st sess. (1852), p. 2.

70. Ellet's analysis first appeared in "Contributions to the Physical Geography of the United States: Of the Physical Geography of the Mississippi Valley," *Smithsonian Contributions to Knowledge* 2 (1851): 1–58; see also Albert Stien, "Mississippi Valley: Remarks on the Improvement of the River Mississippi," *De Bow's Review* 9 (December 1850): 594–601.

71. "Physics and Hydraulics of the Mississippi River," *American Journal of Science and Art,* 33, 2d series (March 1862): 181.

72. Henry H. Humphreys, *Andrew Atkinson Humphreys: A Biography* (Philadelphia: The John C. Winston Co., 1824), p. 52.

73. A. D. Bache to Secretary of War Conrad, April 4, 1850, Letterbooks, 21.48, Andrew Atkinson Humphreys Papers, Historical Society of Pennsylvania, Philadelphia; see also Henry L. Abbot, "Memoir of Andrew Atkinson Humphreys," *Biographical Memoirs,* National Academy of Sciences (1886), 2:203–213, Cullum, *Biographical Register* 1:384–386, and *DAB,* s.v. "Humphreys, Andrew Atkinson."

74. Andrew A. Humphreys and Henry L. Abbot, *Report upon the Physics and Hydraulics of the Mississippi River,* 2d ed. (Washington, D.C.: Gales and Seaton, 1867), p. 162.

75. Charles A. Hartley, *Letter . . . on the Jetties at the Passes of the Mississippi* (Washington, D.C.: National Republican Job Office Printers, 1875), pp. 16–19; for the army theory of levees, see T. A. Lane and E. J. Williams, Jr., "River Hydraulics in 1861," *Journal of the Waterways and Harbor Division,* Proceedings of the American Society of Civil Engineers, 88 (August 1962): 1–12.

76. James D. B. De Bow, "Mr. Calhoun's Report," *Commercial Review,* 2 (September 1846): 84.

77. Richardson, *Messages and Papers of the Presidents* 5:90–91; see also *U.S. Statutes at Large* 10:56–61. Abert promoted the bill by linking harbors to railroads; see "Internal Improvements by Government—Pacific Railroads," *De Bow's Southern and Western Reviews* 12 (March 1852): 402–406.

78. James J. Abert, "Report of the Colonel of Topographical Engineers, December 2, 1853," H. Doc. 1, 33d Cong., 1st sess. (1853), pp. 19–20; see also Regulations in Relation to River and Harbor Improvements, in Correspondence, Reports, Orders, and Regulations, 1851–1853, entry 459, NA, RG 77.

79. George H. Derby, "Memoir of the San Diego River," H. Doc. 1, 33d Cong., 1st sess. (1853), p. 112.

80. John Newton, "Comparison of the relative merits of Owl's Head and Rockland harbors," S. Exec. Doc. 1, 33d Cong., 1st sess. (1853), pp. 230–233; see also John D. Graham, "Annual Report (no. 116) on the harbor improvements of Lake Michigan and St. Clair, December 31, 1855," S. Exec. Doc. 77, 34th Cong., 1st sess. (1856), pp. 20–40.

81. Richardson, *Messages and Papers of the Presidents* 5:259, 261.

82. Ibid., 5:599–607; see also Robert R. Russel, *Improvement of Communication with the Pacific Coast as an Issue in American Politics, 1783–1864* (Cedar Rapids, Iowa:

Torch Press, 1948), pp. 100, 107–108, and Ryan, "War Department Topographical Bureau," p. 305.

83. Larson, *Those Army Engineers,* pp. 81–91; see also Marion J. Klawonn, *The Cradle of the Corps: A History of the New York District, U.S. Army Corps of Engineers, 1775–1975* (n.p.: U.S. Army Corps of Engineers, 1977), pp. 74–75, and Harold Kanareck, *The Mid-Atlantic Engineers: A History of the Baltimore District, U.S. Army Corps of Engineers, 1774–1974* (Washington, D.C.: U.S. Government Printing Office, 1976), pp. 42–44.

84. S. H. Long to J. J. Abert, November 19, 1857, S. Exec. Doc. 11, 35th Cong., 1st sess. (1858), p. 304; see also, Lee S. Dillion, "Locks at Sault Ste. Marie, Michigan," *Military Engineer* 23 (May–June 1931): 205–207, and Irene D. Neu, "The Building of the Sault Canal: 1852–1855," *Mississippi Valley Historical Review* 40 (June 1953): 38–41; for the Weitzel lock, see Charles Moore, ed., *The Saint Marys Falls Canal* (Detroit: Sault Ste. Marie Semi-Centennial Commission, 1907), pp. 138–144.

85. John N. Dickinson, *To Build a Canal: Sault Ste. Marie, 1853–1854 and After* (Columbus: Ohio State University Press, 1981), p. 72.

86. Herman Haupt, *A Consideration of the Plans Proposed for the Improvement of the Ohio River* (Philadelphia: T. K. and P. G. Collins, 1855), p. 52; see also Dickinson, *To Build a Canal,* pp. 39–44, 71; for the civilian complaint against government projects, see *Congressional Globe,* 34th Cong., 1st sess. (1856), p. 2105; see also Calhoun, *American Civil Engineer,* pp. 187–189; for the protest from California, see Sen. John B. Weller's comments on army road building, quoted in Goetzmann, *Army Exploration in the West,* p. 343.

87. Russell F. Weigley, *Quartermaster General of the Union Army: A Biography of M. C. Meigs* (New York: Columbia University Press, 1959), p. 61.

88. M. C. Meigs, "Annual Report of the Operations upon the Washington Aqueduct," S. Exec. Doc. 11, 35th Cong., 1st sess. (1858), pp. 225–282; see also Robert J. Hellman, "The Corps of Engineers U.S. Army and the Water Supply of Washington, D.C," Corps of Engineers History Office (typewritten, 1983), pp. 11–32; and Philip O. MacQueen, "New Aqueduct for the National Capital," *Military Engineer* 27 (1926): 110–117.

89. Harold K. Skramstad, "The Engineer as Architect in Washington: The Contribution of Montgomery Meigs," *Records of the Columbia Historical Society of Washington, D.C.: 1969–1970* (Washington, D.C., Columbia Historical Society, 1970), pp. 268–274; see also Weigley, *Quartermaster General of the Union Army,* pp. 68–73.

90. Sherrod E. East, "The Banishment of Captain Meigs," *Records of the Columbia Historical Society* 40–41 (1940): 130–131.

91. *Congressional Globe,* 36th Cong., 1st sess. (1860), p. 501.

92. East, "Banishment of Captain Meigs," pp. 126, 128–129; see also Richardson, *Messages and Papers of the Presidents* 5:597–599.

93. Weigley, *Quartermaster General of the Union Army,* p. 130.

94. U.S. Army Corps of Engineers, *History of the Washington Aqueduct* (Washington, D.C.: Washington District, Corps of Engineers, 1953), pp. 6, 18; see also Schodek, *Landmarks in American Civil Engineering,* pp. 112–114; for project dimensions, see Silas Seymour, "Annual Report of the Chief Engineer of the Washington Aqueduct, October 1, 1864," H. Exec. Doc. 1, 38th Cong., 2d sess. (1864), pp. 697–713.

95. Raymond H. Merritt, *Engineering in American Society, 1850–1875* (Lexington: University Press of Kentucky, 1969), p. 30; for falling prestige, see Goetzmann, *Army Exploration in the West*, p. 346; see also Jackson, *Wagon Roads West*, pp. 319–328; for Totten, scandals, and the "degradation" of the Corps during the late 1850s, see Kershner, "Sylvanus Thayer," pp. 295–303.

96. Arthur A. Maass, "Congress and Water Resources [September 1950]," in Francis E. Rourke, ed., *Bureaucratic Power in National Politics* (Boston: Little, Brown, & Co., 1972), p. 148.

97. "An Interesting Historical Letter," *Pennsylvania Magazine of History and Biography* 25 (1901): 77–79; see also Weigley, *Quartermaster General of the Union Army*, p. 91.

98. William Baumer, Jr., *Not All Warriors: Portraits of 19th Century West Pointers Who Gained Fame in Other Than Military Fields* (New York: Smith and Durrell, 1941), pp. 226, 229.

99. J. Davis et al, "Report of the Commission . . . to examine . . . the United States Military Academy at West Point," S. Misc. Doc. 3, 36th Cong., 2d. sess. (1860), p. 8.

100. Coffman, *The Old Army*, p. 108.

101. Charlotte Hustler Meade, sister of George Meade, married the topographer Graham. Fort engineer Joseph Mansfield, the father of Corps brigadier general Samuel M. Mansfield, was the nephew of Jared Mansfield, Totten's cousin, and a cousin through marriage to West Pointer Charles Davies and civil engineer Thomas Davies, a builder of the Croton Aqueduct; see *DAB*; see also *National Cyclopedia of American Biography*, 3:26–27, and *Memoirs of General Joseph Swift*, p. 205.

102. Jean Lipman-Blumen, "Toward a Homosocial Theory of Sex Roles: An Explanation of the Sex Segregation of Social Institutions," *Signs* 2 (Spring 1976): 17.

103. Weigley, *Quartermaster General of the Union Army*, p. 109; see also George Meade, *The Life and Letters of George Gordon Meade*, 2 vols. (New York: Charles Scribner's Sons, 1913), 1:17, Sears, *George B. McClellan*, pp. 63, 397; Goetzmann, *Army Exploration in the West*, p. 129, and Dickinson, *To Build a Canal*, p. 39.

104. Martin Van Buren, quoted in *DAB*, s.v. "Macomb, Alexander."

105. Allan Nevins, *Notable American Women, 1607–1950*, s.v. "Frémont, Jessie Ann Benton," ed. Edward T. James (Cambridge: Belknap Press of Harvard University, 1971), 10:668–671.

106. Gustavus W. Smith, "John Newton," *Annual Reunion of the U.S. Military Academy, June 19, 1895* (West Point, N.Y.: Association of Graduates, 1895), p. 108; see also Freeman, *R. E. Lee*, pp. 290–291.

107. Layton, *Revolt of the Engineers*, p. 65; see also Samuel Haber, *The Quest for Authority and Honor in the American Professions, 1750–1900* (Chicago: University of Chicago Press, 1984), p. 296; for gender barriers in science and civil engineering, see Margaret W. Rossiter, *Women Scientists in America: Struggles and Strategies to 1940* (Baltimore: Johns Hopkins University Press, 1982), p. 136, passim.

108. "A New Department of Government," *Iron Age*, February 19, 1885, p. 16.

109. "Civil vs. Military Engineers," *Engineering and Mining Journal*, March 7, 1885, p. 152.

110. For the "backward-looking" engineer bureau, see Philip Lewis Shiman's important study, "Engineering Sherman's March: Army Engineers and the Management of Modern War" (Ph.D. diss., Duke University, 1991), p. 690; see also Allan

Nevins, *The War for the Union*, 4 vols. (New York: Charles Scribner's Sons, 1959), 1:243; for the census statistics, see Merritt, *Engineering in American Society*, p. 10.

111. Shiman, "Engineering Sherman's March," p. 98.

112. Cullum, *Campaigns of the War of 1812*, p. 87; for Meigs's comment, see Weigley, *Quartermaster General of the Union Army*, p. 105.

113. Thian, *Legislative History of the General Staff*, p. 509; see also George T. Ness, Jr., "Engineers of the Civil War," *Military Engineer* 44 (May–June 1952): 179–187, and Ryan, "War Department Topographical Bureau," p. 318.

114. Philip L. Shiman, phone conversation with the author, February 14, 1992.

115. Shiman, "Engineering Sherman's March," p. 327; for Delafield's unpopularity, see also Morrison, *The Best School in the World*, p. 41.

116. Jack Coggins, "The Engineers Played a Key Role in Both Armies," *Civil War Times Illustrated* 3 (January 1965): 40–47; see also U.S. Army Corps of Engineers, Office of History, *The History of the U.S. Army Corps of Engineers* (Washington, D.C.: Corps of Engineers, 1986), pp. 69–71, and Charles W. Larned, "The Genius of West Point," *Centennial of the United States Military Academy* 1:489; for bridge work, see Phillip M. Thienel, "Engineers of the Union Army," *Military Engineer* 47 (January–February 1955): 41; for Haupt's fast-working railroad construction corps, see George Edgar Turner, *Victory Rode the Rails: The Strategic Place of the Railroads in the Civil War* (Indianapolis: Bobbs-Merrill Co., 1953), pp. 208–209.

117. Sen. Daniel W. Voorhees, quoted in Pross, "Rivers and Harbors Appropriation Bills," p. 50; see also Howard B. Schonberger, *Transportation to the Seaboard: The "Communication Revolution" and American Foreign Policy, 1860–1900* (Westport, Conn: Greenwood Publishing Corporation, 1971), pp. 16–32.

118. Thomas G. Manning, *Government in Science: The U.S. Geological Survey, 1867–1894* (Lexington: University of Kentucky Press, 1967), p. 33.

119. Frank D. Carpenter, "Government Engineers," *Lippincott's Magazine* 32 (1883): 159; see also Pross, "Rivers and Harbors Appropriation Bills," p. 71; Kanarek, *Mid-Atlantic Engineers*, p. 70; for Corps irrigation planning, see B. S. Alexander et al., *Engineers and Irrigation: Report of the Board of Commissioners on the Irrigation of the San Joaquin, Tulare, and Sacramento Valleys of the State of California, 1873*, annotated and introduced by W. Turrentine Jackson, Rand F. Herbert, and Stephen R. Wee (Fort Belvoir, Va.: Office of History, 1990).

120. Ambrose, *Duty, Honor, Country*, p. 4; for Corps nepotism see George Y. Wisner, "Worthless Government Engineering," *Engineering Magazine* 2 (January 1892): 434.

121. William E. Merrill, "Improvement of the Ohio River," H. Exec. Doc. 1, 43d Cong., 2d sess. (1874), pp. 400–481; see also Francis H. Oxx, "The Ohio River Movable Dams," *Military Engineer* 27 (January–February, 1935): 49–58, and Leland R. Johnson, *The Davis Island Lock and Dam, 1870–1922* (Pittsburgh: U.S. Engineer District, 1985), pp. 24, 86, 133.

122. Swift to Totten, September 18, 1841, Letters Received, entry 18, NA, RG 77; for origins of Corps levee technology, see Michael C. Robinson, "History of bank protection through the Use of Revetments," in *Centenaire de l'Association Internationale Permanente des Congrès de Navigation, 1885–1985* (Paris: Permanent International Association of Navigation Congresses, 1985), pp. 285–292; Albert E. Cowdrey, *Land's End: A History of the New Orleans District, U.S. Army Corps of Engineers . . .* (New Orleans: U.S. Army Corps of Engineers, 1977), p. 24; and, C. W. S. Hartley,

A Biography of Sir Charles Hartley, Civil Engineer (1825–1915): The Father of the Danube (Lewiston, N.Y.: Edwin Mellen Press, 1989): 317–359.

123. John R. Ferrell, "From Single- to Multi-Purpose Planning: The Role of the Army Engineers in River Development Policy, 1824–1930" (typewritten draft 1976), Historical Division, U.S. Army Corps of Engineers, Baltimore, Maryland, p. 37.

124. C. Vann Woodward, *Reunion and Reaction: The Compromise of 1877 and the End of Reconstruction* (Garden City, N.Y.: Doubleday & Co., 1956), pp. 62, 155–156; see also Gouverneur K. Warren, "Report and Plan for the Reclamation of the Alluvial Basin of the Mississippi River," H. Exec. Doc. 127, 43d Cong., 2d sess. (1875).

125. "The Mississippi," *Saint Paul and Minneapolis Pioneer Press,* October 12, 1877.

126. Roald Tweet, *A History of the Rock Island District, U.S. Army Corps of Engineers* (Rock Island, Ill.: U.S. Army Engineer District, 1984), p. 103; see also Kanarek, *Mid Atlantic Engineers,* p. 61, and George Y. Wisner, "Worthless Government Engineering," *Engineering Magazine* 2 (March 1892): 751.

127. Lowrey, "Navigation Problems at the Mouth of the Mississippi River," pp. 270–301; see also Cowdrey, *Land's End,* p. 18.

128. Charles A. Dana, *Recollections of the Civil War* (New York: Collier Books, 1963), pp. 173–174.

129. John Watts De Peyster, "Andrew Atkinson Humphreys," *Magazine of American History* 16 (October 1886): 351–353; see also Marcy C. Rabbit, *Minerals, Lands, and Geology for the Common Defense and General Welfare,* vol. 1, *Before 1879* (Washington, D.C.: U.S. Government Printing Office, 1979), pp. 216–217; and Manning, *Government in Science,* p. 44.

130. John A. Kouwenhoven, "James Buchanan Eads: The Engineer as Entrepreneur," in Pursell, *Technology in America,* p. 90; see also James B. Eads, *The Mississippi River: Its Hydraulics, Value, and Control* (Washington, D.C.: Ramsey & Bisbee, 1890), pp. 1, 5, and Eads, *Report on the Mississippi Jetties* (New Orleans: n.p., 1876), pp. 19–24; for the Corps' part in the dispute over the Eads Bridge, see Fredrick J. Dobney, *River Engineers on the Middle Mississippi: A History of the St. Louis District, U.S. Army Corps of Engineers* (Washington, D.C.: U.S. Government Printing Office, 1978), pp. 41–43.

131. Charles W. Howell, *Eads' Jetties* (New Orleans: Peter O. Donnell, printer, 1874), p. 3.

132. C. W. Howell, "Fort Saint Phillip Ship-Canal," H. Exec. Doc. 113, 43d Cong., 1st sess. (1874), pp. 36–74; see also Lowrey, "Navigational Problems Mouth of the Mississippi," pp. 413, 422.

133. Lowrey, "Navigational Problems of the Mississippi," p. 432.

134. Woodward, *Reunion and Reaction,* p. 158; for the attempt to discredit Eads, see also Reuss, "Andrew A. Humphreys and the Development of Hydraulic Engineering," p. 23.

135. Max E. Schmidt, "The South Pass Jetties," *Transactions,* American Society of Civil Engineers, 8 (1879): 190–226; see also, Rosemary Yager, *James Buchanan Eads: Master of the Great River* (Princeton, N.J.: D. Van Nostrand, 1968), pp. 90–99.

136. Lowrey, "Navigational Problems of the Mississippi," p. 463.

137. Cyrus B. Comstock, "Discussion on the South Pass of the Jetties," *Transactions,* American Society of Civil Engineers, 15 (April 1886): 232; for Merrill's comments, ibid., pp. 223–224.

138. Morgan, *Dams and Other Disasters,* p. 175.

139. William H. Bixby, "Discussion on the South Pass of the Jetties," p. 255; for New Orleans commerce, see C. Vann Woodward, *Origins of the New South, 1877–1913* (Baton Rouge: Louisiana State University Press, 1951), pp. 125–126.

140. U.S. Army Corps of Engineers, Office of History, *The History of the U.S. Army Corps of Engineers,* p. 114.

141. Ickes, foreword to Maass, *Muddy Waters,* p. ix.

142. Gene Marine, *America the Raped: The Engineering Mentality and the Devastation of a Continent* (New York: Simon and Schuster, 1969), p. 51.

143. Clarke and McCool, *Staking Out the Terrain,* p. 18.

144. Daniel A. Mazmanian and Jeanne Nienaber, *Can Organizations Change?* (Washington, D.C.: The Brookings Institution, 1979), p. 3.

145. Harry Taylor, "Civil Works of the Corps of Engineers," *Military Engineer* 17 (March–April 1925): 95.

146. Franklin M. Davis, Jr., and Thomas T. Jones, *The U.S. Army Engineers—Fighting Elite* (New York: Franklin Watts, Inc., 1967), p. 39.

147. U.S. Senate, Committee on Public Works, "Civil Works Program of the Corps of Engineers," A Report to the Secretary of War by the Civil Works Study Board (Washington, D.C.: U.S. Government Printing Office, 1966), p. 149.

148. Harold C. Livesay, *Andrew Carnegie and the Rise of Big Business* (Boston: Little, Brown and Co., 1975), p. 33; see also Daniel Nelson, *Frederick W. Taylor and the Rise of Scientific Management* (Madison: University of Wisconsin Press, 1980), pp. 154–167; for a bibliographical review, see Barton C. Hacker, "Engineering a New Order: Military Institutions, Technical Education, and the Rise of the Industrial State," *Technology and Culture* 34 (January 1993): 1–27.

149. Joel Mokyr, *The Lever of Riches: Technological Creativity and Economic Progress* (New York: Oxford University Press, 1990), p. 184.

150. Martin Van Creveld, *Technology and War: From 2000 B.C. to the Present* (New York: Free Press, 1989) p. 220.

151. Charles Richard Van Hise, *The Conservation of Natural Resources in the United States* (New York: Macmillan Co., 1927), p. 163; see also George Basalla, *Evolution of Technology* (Cambridge: Cambridge University Press, 1989), p. 196, and David McCullough, *The Path between the Seas: The Creation of the Panama Canal, 1870–1914* (New York: Simon and Schuster, 1977), pp. 532–535, passim.

152. William Ashworth, *Nor Any Drop to Drink* (New York: Summit Books, 1982), p. 125.

153. Reuss, "Andrew Atkinson Humphreys and the Development of Hydraulic Engineering," p. 31; see also Jamie W. Moore and Dorothy P. Moore, *The Army Corps of Engineers and the Evolution of Federal Flood Plan Management Policy* (Boulder: Institute of Behavioral Science, University of Colorado, 1989), pp. 1–15; for resistance to reservoirs, see Morgan, *Dams and Other Disasters,* pp. 252–309.

Epilogue: Formative Conflicts

1. Harry N. Scheiber, "Federalism and the American Economic Order, 1789–1910," *Law and Society Review* 10 (Fall 1975): 58.

2. For the "professional state," see Frederick C. Mosher, *Democracy and the Public Service* (New York: Oxford University Press, 1968), pp. 99–133; for "revolutions" bureaucratic and technocratic, see Don K. Price, *The Scientific Estate* (Cambridge: Harvard University Press, 1965), pp. 15–16; see also Eugene P. Moehring, "Space, Economic Growth, and the Public Works Revolution," in Ann Durkin Keating et al., *Infrastructure and Urban Growth in the Nineteenth Century,* Essays in Public Works History, No. 14 (Chicago: Public Works Historical Society, 1985), pp. 29–59.

3. Florman, "Hired Scapegoats," p. 28.

4. Thorstein Veblen, *The Engineers and the Price System* (New York: Augustus M. Kelley, 1965), p. 151; for the despotic water bureaucracy and the domination of nature, see Donald Worster, *Rivers of Empire: Water, Aridity, and the Growth of the American West* (New York: Pantheon Books, 1985), pp. 284–285, 329.

5. For America's hit list of scapegoats, see Herbert Kaufman, "Fear of Bureaucracy: A Raging Pandemic," *Public Administration Review* 41 (January–February 1981)1 7.

6. E. R. Heiberg III, "Supporting the Army, Strengthening the Nation: The U.S. Army Corps of Engineers," *Constructor* 67 (May 1985): 31.